HAROLD WASHINGTON LIBRARY CENTER

D0850822

Lynn Arthur Steen
J. Arthur Seebach, Jr.

Counterexamples in Topology

Second Edition

Springer-Verlag
New York Heidelberg Berlin

Lynn Arthur Steen
Saint Olaf College
Northfield, Minn. 55057
USA

J. Arthur Seebach, Jr.
Saint Olaf College
Northfield, Minn. 55057
USA

AMS Subject Classification: 54-01

Library of Congress Cataloging in Publication Data

Steen, Lynn A 1941-
Counterexamples in topology.
Bibliography: p.
Includes index.
1. Topological spaces. I. Seebach, J. Arthur, joint author.
II. Title
QA611.3.S74 1978 514′.3 78-1623

The first edition was published in 1970 by Holt, Rinehart, and Winston Inc.

Printed in the United States of America.

9 8 7 6 5 4 3 2 1

ISBN 0-387-90312-7 Springer-Verlag New York

ISBN 3-540-90312-7 Springer-Verlag Berlin Heidelberg

Preface

The creative process of mathematics, both historically and individually, may be described as a counterpoint between theorems and examples. Although it would be hazardous to claim that the creation of significant examples is less demanding than the development of theory, we have discovered that focusing on examples is a particularly expeditious means of involving undergraduate mathematics students in actual research. Not only are examples more concrete than theorems—and thus more accessible—but they cut across individual theories and make it both appropriate and necessary for the student to explore the entire literature in journals as well as texts. Indeed, much of the content of this book was first outlined by undergraduate research teams working with the authors at Saint Olaf College during the summers of 1967 and 1968.

In compiling and editing material for this book, both the authors and their undergraduate assistants realized a substantial increment in topological insight as a direct result of chasing through details of each example. We hope our readers will have a similar experience. Each of the 143 examples in this book provides innumerable concrete illustrations of definitions, theorems, and general methods of proof. There is no better way, for instance, to learn what the definition of metacompactness really means than to try to prove that Niemytzki's tangent disc topology is not metacompact.

The search for counterexamples is as lively and creative an activity as can be found in mathematics research. Topology particularly is replete with unreported or unsolved problems (do you know an example of a Hausdorff topological space which is separable and locally compact, but not σ-compact?), and the process of modifying old examples or creating new ones requires a wild and uninhibited geometric imagination. Far from providing all relevant examples, this book provides a context in which to ask new questions and seek new answers. We hope that each reader will share (and not just vicariously) in the excitement of the hunt.

Counterexamples in Topology was originally designed, not as a text, but as a course supplement and reference work for undergraduate and graduate students of general topology, as well as for their teachers. For such use, the reader should scan the book and stop occasionally for a guided tour of the various examples. The authors have used it in this manner as a supplement to a standard textbook and found it to be a valuable aid.

There are, however, two rather different circumstances under which this monograph could most appropriately be used as the exclusive reference in a topology course. An instructor who wishes to develop his own theory in class lecture may well find the succinct exposition which precedes the examples an appropriate minimal source of definitions and structure. On the other hand, *Counterexamples in Topology* may provide sufficiently few proofs to serve as a basis for an inductive, Moore-type topology course. In either case, the book gives the instructor the flexibility to design his own course, and the students a wealth of historically and mathematically significant examples.

A counterexample, in its most restricted sense, is an example which disproves a famous conjecture. We choose to interpret the word more broadly, particularly since all examples of general topology, especially as viewed by beginning students, stand in contrast to the canon of the real line. So in this sense any example which in some respect stands opposite to the reals is truly a *Gegenbeispiel*. Having said that, we should offer some rationale for our inclusions and omissions. In general we opted for examples which were necessary to distinguish definitions, and for famous, well known, or simply unusual examples even if they exhibited no new properties. Of course, what is well known to others may be unknown to us, so we acknowledge with regret the probable omission of certain deserving examples.

In choosing among competing definitions we generally adopted the strategy of making no unnecessary assumptions. With rare exception therefore, we define all properties for all topological spaces, and not just for, for instance, Hausdorff spaces.

Often we give only a brief outline or hint of a proof; this is intentional, but we caution readers against inferring that we believe the result trivial. Rather, in most cases, we believe the result to be a worthwhile exercise which could be done, using the hint, in a reasonable period of time. Some of the more difficult steps are discussed in the Notes at the end of the book.

The examples are ordered very roughly by their appropriateness to the definitions as set forth in the first section. This is a very crude guide whose only reliable consequence is that the numerical order has no correlation with the difficulty of the example. To aid an instructor in recommending examples for study, we submit the following informal classification by sophistication:

Elementary: 1–25, 27–28, 30–34, 38, 40–47, 49–50, 52–59, 62–64, 73–74, 81, 86–89, 97, 104, 109, 115–123, 132–135, 137, 139–140.

Intermediate: 26, 29, 35–37, 39, 48, 51, 65–72, 75–80, 82–85, 90–91, 93–96, 98–102, 105–108, 113–114, 124, 126–127, 130, 136, 138, 141.

Advanced: 60–61, 92, 103, 110–112, 125, 128–129, 131, 142, 143.

The discussion of each example is geared to its general level: what is proved in detail in an elementary example may be assumed without comment in a more advanced example.

In many ways the most useful part of this book for reference may be the appendices. We have gathered there in tabular form a composite picture of the most significant counterexamples, so a person who is searching for Hausdorff nonregular spaces can easily discover a few. Notes are provided which in addition to serving as a guide to the Bibliography, provide added detail for many results assumed in the first two sections. A collection of problems related to the examples should prove most helpful if the book is used as a text. Many of the problems ask for justification of entries in the various tables where these entries are not explicitly discussed in the example. Many easy problems of the form "justify the assertion that . . ." have not been listed, since these can readily be invented by the instructor according to his own taste.

In most instances, the index includes only the initial (or defining) use of a term. For obvious reasons, no attempt has been made to include in the index all occurrences of a property throughout the book. But the General Reference Chart (pp. 170–179) provides a complete cross-tabulation of examples with properties and should facilitate the quick location of examples of any specific type. The chart was prepared by an IBM 1130 using a program which enables the computer to derive, from the theorems discussed in Part I, the properties for each example which follow logically from those discussed in Part II.

Examples are numbered consecutively and referred to by their numbers in all charts. In those few cases where a minor but inelegant modification of an example is needed to produce the desired concatenation of properties, we use a decimal to indicate a particular point within an example: 23.17 means the 17th point in Example 23.

The research for this book was begun in the summer of 1967 by an undergraduate research group working with the authors under a grant from the National Science Foundation. This work was continued by the authors with support from a grant by the Research Corporation, and again in the summer of 1968 with the assistance of an N.S.F. sponsored undergraduate research group. The students who participated in the undergraduate research groups were John Feroe, Gary Gruenhage, Thomas Leffler, Mary Malcolm, Susan Martens, Linda Ness, Neil Omvedt, Karen Sjoquist, and Gail Tverberg. We acknowledge that theirs was a twofold contribution: not only did they explore and develop many examples, but they proved by their own example the efficacy of examples for the undergraduate study of topology.

Finally, we thank Rebecca Langholz who with precision, forbearance, and unfailing good humor typed in two years three complete preliminary editions of this manuscript.

Northfield, Minn. Lynn Arthur Steen
January 1970 J. Arthur Seebach, Jr.

Preface to the Second Edition

In the eight years since the original edition of *Counterexamples* appeared, many readers have written pointing out errors, filling in gaps in the reference charts, and supplying many answers to the rhetorical question in our preface. In these same eight years research in topology produced many new results on the frontier of metrization theory, set theory, topology and logic.

This *Second Edition* contains corrections to errors in the first edition, reports of recent developments in certain examples with current references, and, most importantly, a revised version of the first author's paper "Conjectures and counterexamples in metrization theory" which appeared in the *American Mathematical Monthly* (Vol. 79, 1972, pp. 113–132). This paper appears as Part III of this *Second Edition* by permission of the Mathematical Association of America.

We would like to thank all who have taken the time to write with corrections and addenda, and especially Eric van Douwen for his extensive notes on the original edition which helped us fill in gaps and correct errors. The interest of such readers and of our new publisher Springer-Verlag has made this second edition possible.

Northfield, Minn. Lynn Arthur Steen
April 1978 J. Arthur Seebach, Jr.

Contents

Part II COUNTEREXAMPLES

PART I
Basic Definitions

SECTION 1
General Introduction

A **topological space** is a pair (X,τ) consisting of a set X and a collection τ of subsets of X, called **open sets**, satisfying the following axioms:

O_1: The union of open sets is an open set.
O_2: The finite intersection of open sets is an open set.
O_3: X and the empty set $\emptyset$ are open sets.

The collection τ is called a **topology** for X. The topological space (X,τ) is sometimes referred to as the **space** X when it is clear which topology X carries.

If τ_1 and τ_2 are topologies for a set X, τ_1 is said to be **coarser** (or **weaker** or **smaller**) than τ_2 if every open set of τ_1 is an open set of τ_2. τ_2 is then said to be **finer** (or **stronger** or **larger**) than τ_1, and the relationship is expressed as $\tau_1 \leq \tau_2$. Of course, as sets of sets, $\tau_1 \subseteq \tau_2$. On a set X, the coarsest topology is the indiscrete topology (Example 4), and the finest topology is the discrete topology (Example 1). The ordering $\leq$ is only a partial ordering, since two topologies may not be **comparable** (Example 8.8).

In a topological space (X,τ), we define a subset of X to be **closed** if its complement is an open set of X, that is, if its complement is an element of τ. The De Morgan laws imply that closed sets, being complements of open sets, have the following properties:

C_1: The intersection of closed sets is a closed set.
C_2: The finite union of closed sets is a closed set.
C_3: X and the empty set $\emptyset$ are both closed.

It is possible that a subset be both open and closed (Example 1), or that a subset be neither open nor closed (Examples 4 and 28).

An F_σ**-set** is a set which can be written as the union of a countable collection of closed sets; a **G_δ-set** is a set which can be written as the intersection of a countable collection of open sets. The complement of every F_σ-set is a G_δ-set and conversely. Since a single set is, trivially, a countable collection of sets, closed sets are F_σ-sets, but not conversely (Example 19). Furthermore, closed sets need not be G_δ-sets (Example 19). By complementation analogous statements hold concerning open sets.

Closely related to the concept of an open set is that of a **neighborhood**. In a space (X,τ), a neighborhood N_A of a set A, where A may be a set consisting of a single point, is any subset of X which contains an open set containing A. (Some authors require that N_A itself be open; we call such sets **open neighborhoods**.) A set which is a neighborhood of each of its points is open since it can be expressed as the union of open sets containing each of its points.

Any collection $\mathcal{S}$ of subsets of X may be used as a **subbasis** (or **subbase**) to generate a topology for X. This is done by taking as open sets of τ all sets which can be formed by the union of finite intersections of sets in $\mathcal{S}$, together with $\varnothing$ and X. If the union of subsets in a subbasis $\mathcal{S}$ is the set X and if each point contained in the intersection of two subbasis elements is also contained in a subbasis element contained in the intersection, $\mathcal{S}$ is called a **basis** (or **base**) for τ. In this case, τ is the collection of all sets which can be written as a union of elements of $\mathcal{S}$. Finite intersections need not be taken first, since each finite intersection is already a union of elements of $\mathcal{S}$. If two bases (or subbases) generate the same topology, they are said to be **equivalent** (Example 28). A **local basis** at the point $x \in X$ is a collection of open neighborhoods of x with the property that every open set containing x contains some set in the collection.

Given a topological space (X,τ), a topology τ_Y can be defined for any subset Y of X by taking as open sets in τ_Y every set which is the intersection of Y and an open set in τ. The pair (Y,τ_Y) is called a **subspace** of (X,τ), and τ_Y is called the **induced** (or **relative**, or **subspace**) **topology** for Y. A set $U \subset Y$ is said to have a particular property **relative to Y** (such as open relative to Y) if U has the property in the subspace (Y,τ_Y). A set Y is said to have a property which has been defined only for topological spaces if it has the property when considered as a subspace. If for a particular property, every subspace has the property whenever a space does, the property is said to be **hereditary**. If every closed subset when considered as a subspace has a property whenever the space has that property, that property is said to be **weakly hereditary**.

An important example of a weakly hereditary property is compactness. A space X is said to be **compact** if from every **open cover**, that is, a collection of open sets whose union contains X, one can select a finite subcollection whose union also contains X. Every closed subset Y of a compact space is compact, since if $\{O_\alpha\}$ is an open cover for Y, $\{O_\alpha\} \cup (X - Y)$ is an open cover for X. From $\{O_\alpha\} \cup (X - Y)$, one can choose a finite subcollection covering X, and from this one can choose an appropriate cover for Y containing only elements of $\{O_\alpha\}$ simply by omitting $X - Y$. A compact subset of a compact space need not be closed (Examples 4, 18).

Limit Points

A point p is a **limit point** of a set A if every open set containing p contains at least one point of A distinct from p. (If the point of A is not required to be distinct from p, p is called an **adherent point**.) Particular kinds of limit points are **ω-accumulation points**, for which every open set containing p must contain infinitely many points of A, and **condensation points**, for which every open set containing p must contain uncountably many points of A. Examples 8 and 32 distinguish these definitions.

The concept of limit point may also be defined for sequences of not necessarily distinct points. A point p is said to be a **limit point of a sequence** $\{x_n\}$, $n = 1, 2, 3, \ldots$ if every open set containing p contains all but finitely many terms of the sequence. The sequence is then said to **converge** to the point p. A weaker condition on p is that every open set containing p contains infinitely many terms of the sequence. In this case, p is called an **accumulation point of the sequence**. It is possible that a sequence has uncountably many limit points (Example 4), both a limit point and an accumulation point that is not a limit point (Example 53), or a single accumulation point that is not a limit point (Example 28).

Since a sequence may be thought of as a special type of ordered set, each sequence has associated with it, in a natural way, the set consisting of its elements. On the other hand, every countably infinite set has associated with it many sequences whose terms are points of the set. There is little relation between the limit points of a sequence and the limit points of its associated set. A point may be a limit point of a sequence, but only an adherent point of the associated set (Example 1). If the points of the sequence are distinct, any accumulation point (and therefore any limit point) of the sequence is an ω-accumulation point of the associated set. Likewise, any ω-accumulation point of a countably infinite set is also an accumulation point (but not necessarily a limit point) of any sequence corresponding to the set. Not too surprisingly, a point may be a limit point of a countably infinite set, but a corresponding sequence may have no limit or accumulation point (Example 8).

If A is a subset of a topological space X, the **derived set** of the set A is the collection of all limit points of A. Generally this includes some points of A and some points of its complement. Any point of A not in the derived set is called an **isolated point** since it must be contained in an open set containing no other point of A. If A contains no isolated points, it is called **dense-in-itself**. If in addition A is closed, it is said to be **perfect**. A closed set A contains all of its limit points since for every $x \in (X - A)$, $X - A$ is an open set containing x and no points of A. Also, a set containing its limit points is closed since $X - A$ contains a neighborhood of each of its points, so is open. Therefore we see that a set is perfect if and only if it equals its derived set.

Closures and Interiors

The **closure** of a set A is the set together with its limit points, denoted by $\bar{A}$ (or A^-). Since a set which contains its limit points is closed, the closure of a set may be defined equivalently as the smallest closed set containing A. Allowing $\bar{A}$ to be A plus its ω-accumulation points or condensation points would permit $\bar{A}$, the closure of A, not to be closed (Example 50.9), which is clearly undesirable. Analogously, we define the **interior** of a set A, denoted by A°, to be the largest open set contained in A, or equivalently, the union of all open sets in A. Clearly the interior of A equals the complement of the closure of the complement of A.

There are at most fourteen different sets that can be formed from a given set A by successive applications of the closure and complement operations. Indeed, these two operations generate a semigroup with fourteen members. These sets are intricately related by inclusion and there is an example of a set A for which all fourteen sets are distinct (Example 32.9). An open set for which $A = A^{-\circ}$ is called **regular open**, and a closed set for which $A = A^{\circ -}$ is called **regular closed**.

The union of the closures of finitely many sets always equals the closure of their union; for infinite collections it need only be contained in the closure of the union (Example 30). Similarly, the intersection of the interiors always contains the interior of the intersection, though they are equal only for finite intersections (Example 32.4). The intersection of finitely many regular open sets is regular open and the union of finitely many regular closed sets is regular closed, but the intersection of regular closed sets need not be regular closed (Example 32.6), and by complementation, the union of regular open sets need not be regular open.

The set of all points which are in the closure of A but not in the interior of A is the **boundary** (or **frontier**) of A, denoted by A^b. A^b is also equal to $A^- \cap (X - A)^-$, since $A^b = A^- - A^\circ = A^- \cap (X - A)^-$. A set is closed if and only if it contains its boundary, and is open if and only if it is disjoint from its boundary. Therefore a set is both open and closed if and only if its boundary is empty. A boundary is always closed since it is the intersection of two closed sets. The boundary of the boundary of a set, A^{bb}, need not equal A^b, although A^{bb} is always contained in A^b (Example 4).

The **exterior** A^e of a set A is the complement of the closure of A, or equivalently, the interior of the complement of A. In general, A° is contained in A^{ee}, but they need not be equal (Example 51). The exterior of the union of sets is always contained in the intersection of the exteriors, and similarly, the exterior of the intersection is contained in the union of the exteriors; equality holds only for finite unions and intersections.

If two sets A and B have the property that $\bar{A} \cap B = A \cap \bar{B} = \varnothing$, they are called **separated**. A set A in a topological space X is **connected** if it cannot be written as the union of two separated sets.

Countability Properties

A set A is said to be **dense** in a space X if every point of X is a point or a limit point of A, that is, if $X = \bar{A}$. A subset A of X is said to be **nowhere dense** in X if no nonempty open set of X is contained in $\bar{A}$. In other words, the interior of the closure of a nowhere dense set is empty. A set is said to be of **first category** (or **meager**) in X if it is the union of a countable collection of nowhere dense subsets of X. Any other set is said to be of **second category**.

A space is said to be **separable** if it has a countable dense subset. It is said to be **second countable** (or **completely separable**, or **perfectly separable**) if it has a countable basis. A space is **first countable** if at each point p of the space, there is a countable local basis, that is, a countable collection of open neighborhoods of p such that each open set containing p contains a member of the collection. Every second countable space is both first countable and separable. The first countability is obvious, while the separability follows from the observation that the union of one point from each basis element forms a countable dense subset. A separable space need not be even first countable (Example 19).

The property of being first countable and the property of being second countable are both hereditary, but the property of being separable is not even weakly hereditary (Example 10). A subspace A of a first countable space is first countable, since the intersection of A with the countable local basis for the space provides a countable local basis for A; similarly, every subspace of a second countable space is second countable.

Functions

Functions on spaces are important tools for studying properties of spaces and for constructing new spaces from previously existing ones. A function f from a space (X,τ) to a space (Y,σ) is said to be **continuous** if the inverse image of every open set is open. This is equivalent to requiring that the inverse image of closed sets be closed, or that for each subset A of X, $f(\bar{A}) \subset \overline{f(A)}$. Another equivalent condition is that for each x in X and each neighborhood N of $f(x)$, there exists a neighborhood M of x such that $f(M) \subset N$. If this last condition holds at a particular point p, the function is said to be **continuous at the point p**.

The composition $g \circ f$ is continuous whenever $f: X \to Y$ and $g: Y \to Z$ are both continuous, since the inverse image under g of an open set in Z is an open set in Y, and the inverse image of that open set under f is again an open set in X.

A function f from (X,τ) to (Y,σ) is said to be **open** if the image under f of each open set is open, and **closed** if the image under f of every closed set is closed. For bijective (one-to-one and onto) functions, the conditions of being open and of being closed are equivalent, although in general they are not equivalent (Example 33). It is not difficult to see that f is an open bijective function if and only if f^{-1} is a continuous bijective function.

A bijective function f from X to Y is a **homeomorphism** if f and f^{-1} are continuous, or equivalently, if f is both continuous and open, or if $f(\bar{A}) = \overline{f(A)}$ for all A. X and Y are then **topologically equivalent** or **homeomorphic**. Such spaces are indistinguishable from a topological point of view. It is possible, though, that two spaces formed by assigning topologies τ and τ^* to a set X may be homeomorphic, even though τ and τ^* are not identical nor even comparable (Example 8.8). It is also possible that two sets, A and B, $A \subset X$ and $B \subset Y$ where X and Y are homeomorphic, may be topologically equivalent as subspaces, but because of the nature of X and Y there may be no homeomorphism of X and Y taking A onto B (Example 32.7).

A property is said to be a **topological invariant** (or **topological property**) if whenever one space possesses a given property, any space homeomorphic to it also possesses the same property. Similarly, a property is called a **continuous**, **open**, or **closed invariant** if any continuous (respectively open, closed) image of a space possessing the property also possesses the property. Both separability and compactness are continuous invariants.

For a given collection of topological spaces (X_α,τ_α), where $\alpha \in A$, an indexing set, the **product space** is defined to be the usual Cartesian product ΠX_α of all the sets X_α, together with the coarsest topology on this set such that all of the coordinate projections π_α are continuous. This coarsest topology is called the Tychonoff topology and has as a subbasis all inverse images under projections of open sets of the X_α's, that is, "open cylinders" of the form $\pi_\alpha^{-1}(U)$. It follows immediately from this description of the subbasis that $f: Z \to \Pi X_\alpha$ is continuous iff $\pi_\alpha \circ f$ is continuous for each α.

If (X,τ) is a topological space, Y a set, and $f: X \to Y$ a function, there is then a finest topology σ for Y relative to which f is continuous. We may describe σ explicitly by noting that $V \subset Y$ is an element of σ (open in Y) iff $f^{-1}(V)$ is in τ. This topology, which depends on $f: X \to Y$ and τ, is called the **identification topology** on Y with respect to f and (X,τ). Now if R is an equivalence relation on X, if $\rho: X \to X/R$ is the usual projection function which maps each $x \in X$ to its equivalence class $[x]$ in X/R, and if σ is the identification topology on X/R with respect to ρ, then $(X/R,\sigma)$ is called the **quotient space** of (X,τ) by the relation R. An important special case arises whenever A is a subspace of X. One may then define an equivalence relation R on X by declaring $x \sim y$ iff $x = y$ or x and y are both in A. In this case X/R is usually written X/A and is called the quotient of X by A.

If (X,τ) and (Y,σ) are two topological spaces, the **topological sum** (Z,φ) of X and Y is defined by taking for the set Z the disjoint union of the sets X and Y, that is, the union of X and Y where X and Y are decreed to have no common elements. The topology φ is defined as the topology generated by the union of τ and σ. φ is characterized by being the finest topology on Z in which the inclusion functions from (X,τ) and (Y,σ) are continuous.

Filters

A **filter** on a set X is a collection F of subsets of X with the following properties:

- F_1: Every subset of X which contains a set of F belongs to F.
- F_2: Every finite intersection of sets of F belongs to F.
- F_3: The empty set is not in F.

The set X with the filter F is called a set **filtered by F**, or just a **filtered set**.

If $\mathcal{B}$ is a nonempty set of subsets of X which does not contain $\emptyset$, then the collection of all subsets of X which contain some member of $\mathcal{B}$ is a filter F if and only if the intersection of any two sets in $\mathcal{B}$ contains a set in $\mathcal{B}$. Such a set $\mathcal{B}$ is called a **base** of the filter F and F is called the filter generated by $\mathcal{B}$. Equivalently, a subset $\mathcal{B}$ of a filter F is a base of F if and only if every set of F contains a set of $\mathcal{B}$. Two filter bases are said to be equivalent if they generate the same filter. Conditions F_2 and F_3 imply that the family of sets F satisfies the **finite intersection property**, that is, that the intersection of any finite number of sets of the family is nonempty. Conversely, any family of sets satisfying the finite intersection property is a **subbase** for a filter F since the family together with the finite intersections of its members is a filter base.

If F, F' are two filters on the same set X, F' is said to be **finer** than F

(or F is **coarser** than F') if $F \subset F'$. If also $F \neq F'$, then F' is said to be **strictly finer** than F, or F **strictly coarser** than F'. Two filters are said to be **comparable** if one is finer than the other. A filter F' with base $\mathcal{B}'$ is finer than a filter F with a base $\mathcal{B}$ if and only if every set of $\mathcal{B}$ contains a set of $\mathcal{B}'$.

If a filter F on X has the property that there is no filter on X which is strictly finer than F, F is called an **ultrafilter** on X. Equivalently, F is an ultrafilter if and only if for every two disjoint subsets A and B of X such that $A \cup B \in F$, then either $A \in F$ or $B \in F$. Thus if F is an ultrafilter and $E \subset X$ then either E or $X - E$ is in F. Furthermore, if F and F' are distinct, there exists a set A such that $A \in F$ and $A \notin F'$; but then $X - A \in F'$, so we have $A \in F$ and $X - A \in F'$.

If a point x is in all the sets of a filter we call it a **cluster point**; clearly an ultrafilter can have at most one cluster point. An ultrafilter with a cluster point p is just the set of all sets containing that point and is called a **fixed**, or **principal ultrafilter**; an ultrafilter with no cluster point is called **free**, or **nonprincipal**.

If X is a topological space, the set N of all neighborhoods of an arbitrary nonempty subset A of X is called the **neighborhood filter of** $\boldsymbol{A}$. Let F be any filter on X. A point $x \in X$ is said to be a **limit point** of F if F is finer than the neighborhood filter N of x; F is also said to **converge** to x. The point x is said to be a **limit of a filter base** $\mathcal{B}$ on X, and $\mathcal{B}$ is said to **converge** to x, if the filter whose base is $\mathcal{B}$ converges to x. Equivalently, a filter base $\mathcal{B}$ on a topological space X is said to converge to x if and only if every neighborhood of x contains a set of $\mathcal{B}$.

SECTION 2

Separation Axioms

It is often desirable for a topologist to be able to assign to a set of objects a topology about which he knows a great deal in advance. This can be done by stipulating that the topology must satisfy axioms in addition to those generally required of topological spaces.

One such collection of conditions is given by means of axioms called T_i or separation axioms. These stipulate the degree to which distinct points or closed sets may be **separated by open sets**. Let (X,τ) be a topological space.

T_0 axiom: If $a,b \in X$, there exists an open set $O \in \tau$ such that either $a \in O$ and $b \notin O$, or $b \in O$ and $a \notin O$.

T_1 axiom: If $a,b \in X$, there exist open sets O_a, $O_b \in \tau$ containing a and b respectively, such that $b \notin O_a$ and $a \notin O_b$.

T_2 axiom: If $a,b \in X$, there exist disjoint open sets O_a and O_b containing a and b respectively.

T_3 axiom: If A is a closed set and b is a point not in A, there exist disjoint open sets O_A and O_b containing A and b respectively.

T_4 axiom: If A and B are disjoint closed sets in X, there exist disjoint open sets O_A and O_B containing A and B respectively.

T_5 axiom: If A and B are separated sets in X, there exist disjoint open sets O_A and O_B containing A and B respectively.

If (X,τ) satisfies a T_i axiom, X is called a **T_i space**. A T_0 space is sometimes called a **Kolmogorov space** and a T_1 space, a **Fréchet space**. We will conform to common practice and call a T_2 space a **Hausdorff space**.

It follows from the T_i axioms that T_0 spaces are characterized by the fact that no two points can be limit points of each other. Similarly, T_1 spaces are characterized by points being closed, and T_2 spaces by points being the intersection of their closed neighborhoods. T_3 spaces may be characterized either by the fact that each open set contains a closed neighborhood around each of its points, or by the property that each closed set is the intersection of its closed neighborhoods. A space is T_4 iff every open set O contains a closed neighborhood of each closed set contained in O. It is T_5 iff every subset Y contains a closed neighborhood of each set $A \subset Y^\circ$ where $\bar{A} \subset Y$.

Each of these axioms is independent of the axioms for a topological space; in fact there exist examples of topological spaces which fail to satisfy any T_i axiom (Example 21). But they are not independent of each other, since for instance, axiom T_2 implies axiom T_1 and axiom T_1 implies axiom T_0. There are, on the other hand, T_0 spaces which fail to satisfy every other separation axiom (Example 53) and T_1 spaces which do not satisfy any separation axiom but the T_0 axiom (Example 18); similarly, there are T_2 spaces which fail to be T_3, T_4 or T_5 (Example 75). Furthermore, neither the T_3 axiom nor the T_4 axiom implies any of the other separation axioms (Examples 90.4 and 21.8) nor is either generally implied by them though in compact spaces T_2 implies T_4 but not T_5 (Example 86). The T_5 axiom does however imply T_4, though it is independent of the other separation axioms.

Regular and Normal Spaces

More important than the separation axioms themselves is the fact that they can be employed to define successively stronger properties. To this end, we note that if a space is both T_3 and T_0, it is T_2, while a space that is both T_4 and T_1 must be T_3. The former spaces are called regular, and the latter normal.

Specifically a space X is said to be **regular** if and only if it is both a T_0 and a T_3 space; to be **normal** if and only if it is both a T_1 and a T_4 space; to be **completely normal** if and only if it is both a T_1 and a T_5 space. Thus, we have the following sequence of implications:

$$\text{Completely normal} \Rightarrow \text{Normal} \Rightarrow \text{Regular} \Rightarrow \text{Hausdorff} \Rightarrow T_1 \Rightarrow T_0.$$

Examples 86, 82, 75, 18, and 53 show that the implications are not reversible. A T_4 space (or T_5 space) must be a T_1 space in order to guarantee that it is a T_3 space, for there are T_4 spaces (and T_5 spaces) which are T_0 and yet fail to be T_3 (Example 55).

The use of the terms "regular" and "normal" is not uniform throughout the literature. While some authors use these terms interchangeably with "T_3 space" and "T_4 space" respectively, others refer to our T_3 space as a "regular" space and vice versa, and similarly permute "T_4 space" and "normal." This allows the successively stronger properties to correspond to increasing T_i axioms. We prefer, however, to allow a T_i space to be a space satisfying the corresponding T_i axiom, and content ourselves with labeling the successively stronger properties with unique terminology.

Completely Hausdorff Spaces

We will now introduce two variations of the separation properties. The first involves the use of closed neighborhoods in place of open sets in axioms T_2, T_3, and T_4.

Since in normal spaces every open set O contains a closed neighborhood of each closed set contained in O, if X is a normal space and if A and B are disjoint closed subsets, there exist open sets O_A and O_B containing A and B, respectively, such that $\bar{O}_A \cap \bar{O}_B = \varnothing$. So the use of closed neighborhoods in place of open sets in the definition of a normal space yields the same class of spaces.

Similarly, if X is a regular space, A is a closed subset, and b is a point not in A, then there are open sets O_A and O_b containing A and b respectively such that $\bar{O}_A \cap \bar{O}_b = \varnothing$. However, there are Hausdorff spaces which have two points which do not have disjoint closed neighborhoods (Example 75). Thus, we present the following new axiom.

$T_{2\frac{1}{2}}$ axiom: If a and b are two points of a topological space X, there exist open sets O_a and O_b containing a and b, respectively, such that $\bar{O}_a \cap \bar{O}_b = \varnothing$.

A $T_{2\frac{1}{2}}$ space will be called a **completely Hausdorff space.** It is clear that every regular space is completely Hausdorff and every completely Hausdorff space is Hausdorff. Since there are completely Hausdorff spaces which fail to be regular (Example 78), the completely Hausdorff property is intermediate in strength between the properties Hausdorff and regular.

Completely Regular Spaces

The second variation of the separation axioms concerns the existence of certain continuous, real-valued functions. A **Urysohn function** for A and B, disjoint subsets of a space X, is a continuous function $f\colon X \to [0,1]$ such that $f|_A = 0$ and $f|_B = 1$.

Urysohn's famous lemma asserts that if A and B are disjoint closed subsets of a T_4 space, there exists a Urysohn function for A and B. Conversely, if there is a Urysohn function for any two disjoint closed sets A and B in a space X, then X is a T_4 space. But the existence of such a function does not guarantee that X is a T_1 space and thus a normal space (Example 5).

However, the statement for regular spaces analogous to Urysohn's lemma is false (Example 90), so we give the following new separation axiom:

$T_{3\frac{1}{2}}$ axiom: If A is a closed subset of a space X and b is a point not in A, there is a Urysohn function for A and $\{b\}$.

Then every $T_{3\frac{1}{2}}$ space is a T_3 space, though not necessarily a T_2 space (Example 5) unless it is also a T_0 space. Such a space, which is both T_0 and $T_{3\frac{1}{2}}$, will be called **completely regular**, or **Tychonoff**. Thus completely regular spaces are regular, Hausdorff, and therefore T_1. Since points are closed in normal spaces, it follows from Urysohn's lemma that normal spaces are completely regular. There are, however, regular spaces which fail to be completely regular (Example 90) and completely regular spaces which fail to be normal (Example 82).

Although normal spaces are $T_{3\frac{1}{2}}$, T_4 spaces need not be (Example 55). But if a T_4 space is also T_3, even though possibly not normal, it must nevertheless be $T_{3\frac{1}{2}}$ for if the point p is disjoint from the closed set A in a T_3 space, then there is an open neighborhood of A disjoint from p. The complement B of this neighborhood is a closed set disjoint from A, so if the space is also T_4, we can apply Urysohn's lemma to produce a Urysohn function for A and B. This function is clearly a Urysohn function for A and $\{p\}$ also.

We summarize the T_i implications and counterexamples in Figure 1 where the numbers in parentheses indicate examples. In addition, we have the following simple diagram, in which none of the arrows reverse, and where the numbers refer to appropriate counterexamples:

$$\text{Comp. norm.} \underset{(86)}{\Rightarrow} \text{Normal} \underset{(82)}{\Rightarrow} \text{Comp. reg.} \underset{(90)}{\Rightarrow} \text{Regular} \underset{(78)}{\Rightarrow} \text{Comp. Haus.} \underset{(75)}{\Rightarrow} T_2 \underset{(18)}{\Rightarrow} T_1 \underset{(8)}{\Rightarrow} T_0.$$

Functions, Products, and Subspaces

All the separation properties are topological properties, that is, they are preserved under homeomorphisms. However, certain of the properties are preserved under less restrictive functions.

If X and Y are topological spaces, and $f: X \to Y$ is a closed bijection, and X is T_0, T_1, Hausdorff, or completely Hausdorff, then Y is T_0, T_1, Hausdorff, or completely Hausdorff, respectively. In particular, if $\tau_1 \subset \tau_2$ are topologies for X, the identity function from (X,τ_1) to (X,τ_2) is closed; hence, if (X,τ_1) is T_0, T_1, Hausdorff, or completely Hausdorff, then so is

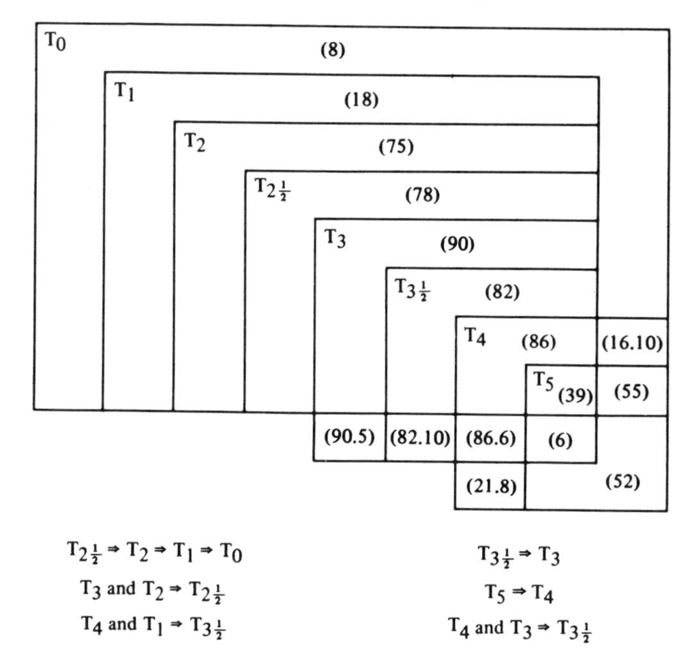

Figure 1.

(X,τ_2). We call τ_2 an **expansion** of τ_1 and note that expansion preserves the above separation properties. The stronger separation properties are not, in general, preserved under expansion (Example 66).

Most separation properties are, however, preserved under products. If $X = \Pi X_\alpha$, then X is a T_0, T_1, Hausdorff, completely Hausdorff, regular, or completely regular space if and only if each of the X_α is T_0, T_1, Hausdorff, completely Hausdorff, regular, or completely regular, respectively. If X is normal or completely normal, each X_α is normal or completely normal, but the converse does not hold (Example 84).

Normality diverges from the remainder of the separation properties in the case of subspaces, also. For every subspace of a T_0, T_1, Hausdorff, completely Hausdorff, regular, or completely regular space is T_0, T_1, Hausdorff, completely Hausdorff, regular, or completely regular, respectively. But only closed subspaces of normal spaces need be normal (Example 86). However, every subspace of a completely normal space is completely normal, since, a space is completely normal iff every subspace is normal. In fact, a space is T_5 iff every subspace is T_4.

Additional Separation Properties

Urysohn's lemma guarantees for T_4 spaces the existence of a Urysohn function for any two disjoint closed sets. Requiring such a function for a point and a closed set gave the $T_{3\frac{1}{2}}$ property which was stronger than axiom T_3. When applied to two points this requirement yields a condition even stronger than completely Hausdorff (Example 80). We call a space with a Urysohn function for any two points a **Urysohn space**.

A T_4 space in which every closed set is a G_δ is often called **perfectly T_4**. A perfectly T_4 space which is also T_1 will be called **perfectly normal**. Every perfectly normal space is completely normal, but not conversely (Example 24).

Since each open set in a T_3 space contains a closed neighborhood around each of its points, every open set in a T_3 space can be written as the union of regular open sets. Since the converse is not true (Example 81), we will call **semiregular** all T_2 spaces in which the regular open sets form a basis for the topology. Semiregular spaces are not necessarily either completely Hausdorff (Example 81) or Urysohn (Example 80).

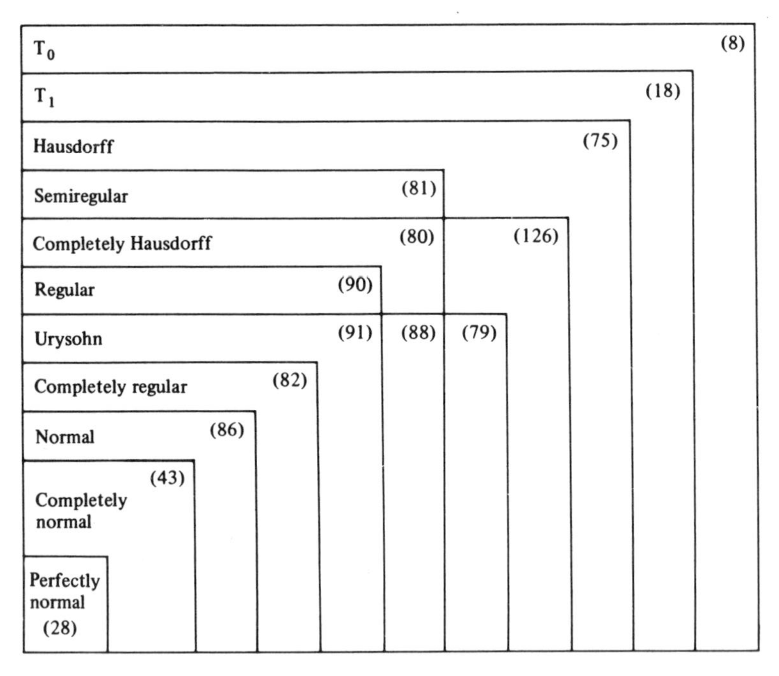

Figure 2.

If we add these new properties to the basic separation axiom structure, we obtain the summary Figure 2. The implications are best illustrated by

$$\begin{array}{ccccccccc}
\begin{array}{c}\text{Perf.}\\ \text{norm.}\end{array} & \Rightarrow & \begin{array}{c}\text{Comp.}\\ \text{norm.}\end{array} & \Rightarrow & \text{Normal} & \Rightarrow & \begin{array}{c}\text{Comp.}\\ \text{reg.}\end{array} & \Rightarrow & \text{Urysohn} \\
 & & & & & & \Downarrow & & \Downarrow \\
 & & & & & & \text{Regular} & \Rightarrow & \begin{array}{c}\text{Comp.}\\ \text{Haus.}\end{array} \\
 & & & & & & \Downarrow & & \Downarrow \\
 & & & & & & \begin{array}{c}\text{Semi-}\\ \text{regular}\end{array} & \Rightarrow & T_2 \Rightarrow T_1 \Rightarrow T_0.
\end{array}$$

SECTION 3

Compactness

A space satisfies a certain separation axiom only if the topology contains enough open sets to provide disjoint neighborhoods for certain disjoint sets. Compactness, however, limits the number of open sets in a topology, for every open cover of a compact topological space must contain a finite subcover. This difference between the separation axioms and the various forms of compactness is illustrated in the extreme by the double pointed finite complement topology (Example 18.7) which is not even T_0 yet does satisfy all the forms of compactness.

Global Compactness Properties

A topological space X is **compact** if every open cover contains a finite subcover; equivalently, X is compact if it satisfies the **finite intersection axiom**, that is, if every family of closed subsets whose intersection is empty contains a finite subfamily whose intersection is empty. For if $\{A_\alpha\}$ is any family of closed sets such that $\bigcap A_\alpha = \varnothing$, then $\{X - A_\alpha\}$ is an open cover which has a finite subcover $\{X - A_{\alpha_k} | k \leq n\}$. By De Morgan's Law, $X - \bigcup(X - A_{\alpha_k}) = \varnothing$ iff $\bigcap A_{\alpha_k} = \varnothing$. Conversely, if the family $\{O_\alpha\}$ is an open cover of X, then since $\bigcap(X - O_\alpha) = \varnothing$, there is a finite subfamily such that $\bigcap_{k=1}^{n} (X - O_{\alpha_k}) = \varnothing$. By De Morgan's Law, $\bigcup_{k=1}^{n} O_{\alpha_k} = X$. An equivalent subbasis condition for compactness is given by Alexander's Compactness Theorem: if a topological space X has a subbasis S such that from every cover of X by elements of S, a finite subcover can be selected, then X is compact. The condition is clearly necessary, but the proof of sufficiency uses the axiom of choice.

Two generalizations of compactness may be obtained by weakening the requirement that subcovers be finite. A topological space is called **σ-compact** if it is the union of countably many compact sets, while a space is called **Lindelöf** if every open cover has a countable subcover. Clearly every compact space is σ-compact and every σ-compact space Lindelöf. These implications are not reversible (Examples 28 and 51).

A topological space is called **countably compact** if any one of the following equivalent conditions is satisfied:

- CC_1: Every countable open cover of X has a finite subcover.
- CC_2: Every infinite set has an ω-accumulation point in X.
- CC_3: Every sequence has an accumulation point in X.
- CC_4: Every countable collection of closed sets with an empty intersection has a finite subfamily with an empty intersection.

Condition CC_4, the countable finite intersection axiom, is equivalent to CC_1 for the same reasons that the ordinary finite intersection axiom is equivalent to compactness. Conditions CC_2 and CC_3 are equivalent to each other since a point is an ω-accumulation point of a countably infinite set iff it is an accumulation point of that set viewed as a sequence. Now if the space X has a countable open cover $\{O_i\}$ with no finite subcover, we can find a set $\{x_n\}$ of distinct points such that $x_n \notin \bigcup_{i=1}^{n} O_i$; this sequence can have no ω-accumulation point in X, for every point of X has a neighborhood, namely one of the O_i to which it belongs, which intersects only finitely many points of the set. Thus $CC_2 \Rightarrow CC_1$. Conversely, if $S \subset X$ is a countably infinite set without an ω-accumulation point, each $x \in X$ would have an open neighborhood O_x which intersects at most finitely many points of S. For each finite subset F of S, define $O_F = \bigcup\{O_x | O_x \cap S = F\}$. Then $\{O_F\}$ is a countable open covering of X every finite subcollection of which includes at most finitely many points of S. Thus no finite subcollection may cover X.

Two other conditions are closely related, but not equivalent to countable compactness: a topological space is said to be **sequentially compact** if every sequence has a convergent subsequence, and **weakly countably compact** (or **limit point compact**) if every infinite set has a limit point. Sequential compactness clearly implies countable compactness, and since every ω-accumulation point is a limit point, every countably compact space is weakly countably compact. However, neither converse is necessarily true (Examples 105 and 106). However, in a T_1 space, weak countable compactness is equivalent to countable compactness. For assuming that x is a limit point of a set A, but not an ω-accumulation point, implies that some open set O_x containing x contains only a finite number of points of A, say $\{a_1 \ldots a_n\}$. But in a T_1 space, this implies that x has an open neighborhood which contains no points of A, that is, that x is not a limit point of A.

Finally, a space X is called **pseudocompact** if every continuous real-valued function on X is bounded. Every countably compact space X is pseudocompact, since for a continuous function f on X, the sets $S_n = \{x \mid |f(x)| < n\}$ form a countable cover of X whose finite subcover yields a bound for the absolute value of f. Although the converse is not true (Example 9), every pseudocompact normal space is necessarily countably compact. Suppose not; then X would contain an infinite subset $S = \{x_n\}$ with no ω-accumulation point. Since X is T_1, S is closed and discrete in the subspace topology; since X is T_4, the Tietze extension theorem guarantees a continuous extension to X of the unbounded continuous function $f: S \to R$ defined by $f(x_n) = n$. This shows that X could not have been pseudocompact.

The relations between the varieties of global compactness may be summarized in this diagram:

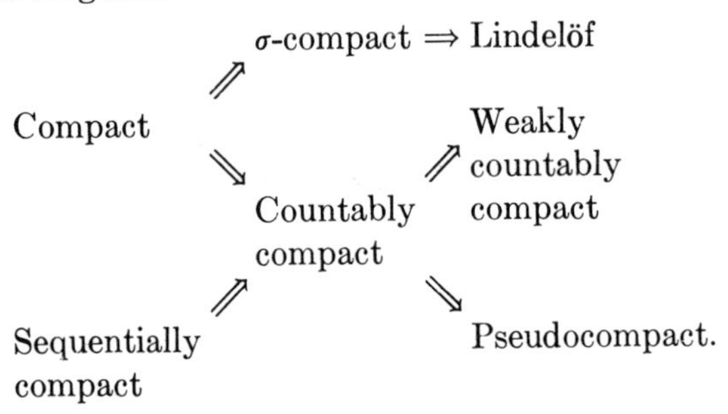

In general, none of the arrows reverse, though, trivially, every countably compact Lindelöf space is compact. So Figure 3 summarizes both the implications and counterexamples.

Localized Compactness Properties

A topological space is called **locally compact** if each point is contained in a compact neighborhood. Clearly every compact space X is locally compact, since X itself is a compact neighborhood of each of its points.

A common nonequivalent variation of the definition of local compactness requires that each point be contained in an open set whose closure is compact. We shall call this concept **strong local compactness** since every space satisfying this condition is clearly locally compact; the converse, however, is not generally true (Example 52) although it does hold in Hausdorff spaces for in such spaces compact sets are closed, so the interior of every compact neighborhood has a compact closure.

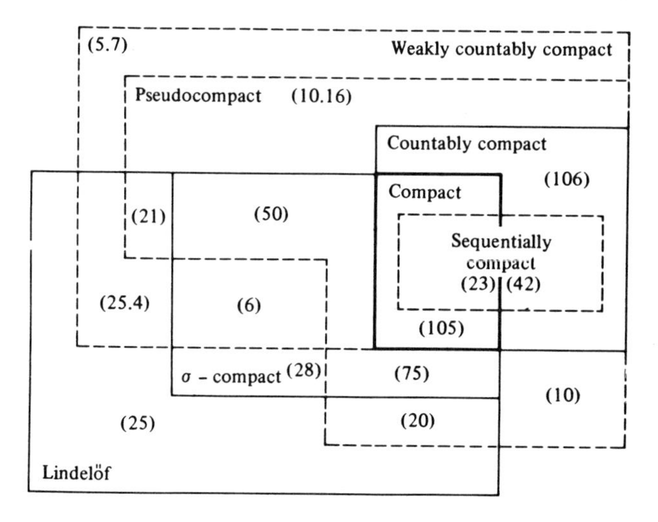

Figure 3.

A different strong form of local compactness is obtained by requiring X to be both σ-compact and locally compact: such a space is called **σ-locally compact**. It suffices in fact to assume X locally compact and Lindelöf, for such spaces must be σ-compact: the interiors of the compact neighborhoods cover X, so some countable number of such interiors, and therefore of compact neighborhoods, covers X.

Although both stronger properties imply local compactness, strong local compactness and σ-local compactness are independent (Examples 3 and 52). We may summarize the implications as follows:

Compact	$\Rightarrow$	σ-locally compact	$\Rightarrow$	σ-compact	$\Rightarrow$	Lindelöf
$\Downarrow$		$\Downarrow$				
Strongly locally compact	$\Rightarrow$	Locally compact.				

The appropriate counterexamples are summarized in Figure 4.

Countability Axioms and Separability

Although the previous compactness properties indirectly imply limitations on the number of open sets in a topology, the countability axioms introduced in the first section directly limit the number of open sets by restricting the number of basis elements. There are three major countability properties: a topological space is **separable** if it has a countable dense

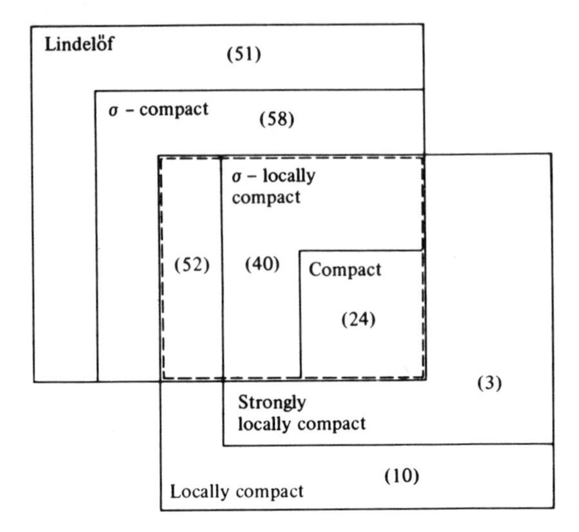

Figure 4.

subset, **second countable** if it has a countable basis, and **first countable** if the neighborhood system of every point has a countable local basis. Clearly, every second countable space is first countable, separable and Lindelöf, although none of these implications reverse. In fact, there are spaces which are first countable, separable and Lindelöf but not second countable (Example 51). A special property which is strictly weaker than separability (Example 20) is the **countable chain condition**, which is the condition that every disjoint family of open sets is countable.

In second countable spaces, compactness is equivalent to countable compactness. Similarly, in a first countable space, countable compactness is equivalent to sequential compactness, for if $\{s_n\}$ is any sequence in a countably compact space X with accumulation point $p \in X$, there is a countable local base at p, say $\{V_n | V_1 \supset V_2 \supset V_3 \ldots\}$. Then a subsequence $\{s_{n_i}\}$ where $s_{n_i} \in V_i$ converges to p.

Figure 5 summarizes the important relations between the countability axioms and compactness.

Paracompactness

Several compactness properties which have both local and global aspects rely on the concept of a refinement of a cover. A cover $\{V_\beta\}$ of a space X is a **refinement** of a cover $\{U_\alpha\}$ if for each V_β there is a U_α such that $V_\beta \subset U_\alpha$. A cover is **point finite** if each point belongs to only finitely many sets in the covering, and it is **locally finite** if each point has some neighborhood which intersects only finitely many members of the cover.

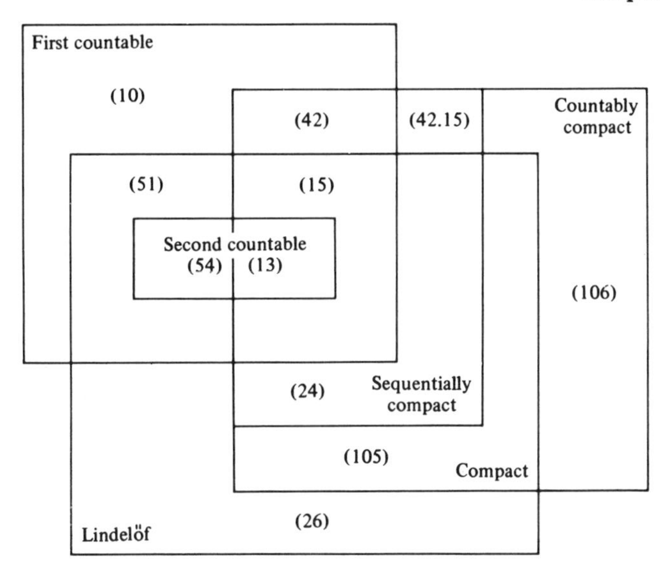

Figure 5.

Finally, a cover $\{V_\beta\}$ of X is said to be a **star refinement** of a cover $\{U_\alpha\}$ if for each $x \in X$ there is some U_α such that $x^* \subset U_\alpha$, where x^*, the **star** of x with respect to $\{V_\beta\}$, is the union of all the sets V_β of which x is an element.

A space is called **metacompact** (or sometimes **pointwise paracompact**) if every open cover has an open point finite refinement, **paracompact** if every open cover has an open locally finite refinement, and **fully T_4** if every open cover has a star refinement. The slightly weaker conditions of **countable metacompactness** and **countable paracompactness** require only that every countable open cover have the desired type of refinement. A fully T_4 space which is also T_1 is called **fully normal.** As the notation implies, every fully normal space is normal, and also paracompact.

Clearly every compact space is paracompact, and every paracompact space metacompact. Although these implications are not reversible (Examples 28 and 89), every metacompact space (and therefore every paracompact space) which is also countably compact must be compact. For if $\{U_\alpha\}$ is any open covering of the metacompact space X, $\{U_\alpha\}$ has an open point finite refinement $\{V_\beta\}$. Now $\{V_\beta\}$ has an irreducible (that is, a minimal) subcovering $\{V_\gamma\}$, for if we order subcoverings by inclusion, the intersection of a chain of subcoverings is a subcovering: if x is not covered by the intersection of the subcoverings, being contained in only finitely many V_β, it would fail to be covered by one of the elements of the chain of subcoverings, a contradiction. Now $\{V_\gamma\}$ is a finite covering, for in each V_γ there is an

x_γ belonging to no other element of the family $\{V_\gamma\}$ since the family is minimal and if the family $\{V_\gamma\}$ were infinite the set $\{x_\gamma\}$ would be an infinite set with no ω-accumulation point.

Thus we have the following implications:

$$\begin{array}{ccccc}
 & & \text{Fully normal} & \Rightarrow & \text{Fully } T_4 \Rightarrow T_4 \\
 & & \Downarrow & & \\
\text{Compact} & \Rightarrow & \text{Paracompact} & \Rightarrow & \text{Metacompact} \\
\Downarrow & & \Downarrow & & \Downarrow \\
\text{Countably compact} & \Rightarrow & \text{Countably paracompact} & \Rightarrow & \text{Countably metacompact.}
\end{array}$$

None of the implications is reversible, so Figure 6 can be used to summarize the necessary counterexamples.

Just as a Lindelöf countably compact space is compact every Lindelöf countably metacompact space is metacompact and every Lindelöf countably paracompact space is paracompact. Furthermore a separable metacompact space is Lindelöf. For if $\{U_\alpha\}$ is an open cover with no countable subcover, and $\{V_\beta\}$ is a point finite refinement (uncountable, of course), and $\{x_\delta\}$ is a countable dense subset, then each V_β contains some x_δ so some x_δ is contained in uncountably many V_α, a contradiction to the nature of $\{V_\beta\}$.

Compactness Properties and the T_i Axioms

Although compactness and the separation axioms involve conflicting requirements on the number of open sets in the topology, when compactness properties are combined with the T_2 or T_3 axioms, the topology often

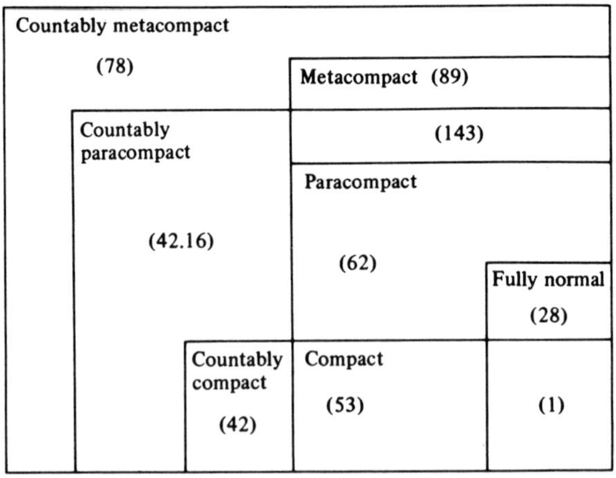

Countably metacompact (78)
Metacompact (89)
(143)
Countably paracompact (42.16)
Paracompact (62)
Fully normal (28)
Countably compact (42)
Compact (53)
(1)

Figure 6.

satisfies certain higher T_i axioms. For the compactness properties, by limiting the number of open sets in a cover, allow the desired disjoint open sets to be constructed by finite intersections. As a result compact sets in T_2 or T_3 spaces have the same properties as points, namely: two disjoint compact sets in a Hausdorff space have disjoint neighborhoods, while if A is a compact subset of T_3 spaces, then for each open set $U \supset A$, there is an open V such that $A \subset V \subset \bar{V} \subset U$. Also, then, compact sets in a Hausdorff space are closed. Thus a compact Hausdorff space is T_4 since closed subsets of a compact space are compact. In fact certain conditions weaker than compactness are sufficient for this, while other compactness properties result in only weaker conclusions. The precise nature of the implications following from the assumption of the T_2 or T_3 separation axioms is pictured in Figure 7. Furthermore, certain combinations of the compactness properties and separation axioms force a space to be of the second category in itself. This type of Baire category theorem applies both to locally compact Hausdorff spaces and to countably compact regular spaces.

Compact Hausdorff topologies are especially interesting since any such topology τ on a space X is both minimal Hausdorff as well as maximal compact. τ is a minimal Hausdorff topology since if $\tau^* \subset \tau$, the identity map $f: (X,\tau) \to (X,\tau^*)$ would be continuous. Thus if A is closed in (X,τ), it is compact (since (X,τ) is compact) and thus $f(A)$ is compact. If τ^* were Hausdorff, $f(A)$ would be closed, and hence f would be a closed mapping—which would mean that $\tau \subset \tau^*$. Thus no topology strictly smaller than τ can be Hausdorff. Similarly, τ is a maximal compact topology for if $\tau^* \supset \tau$,

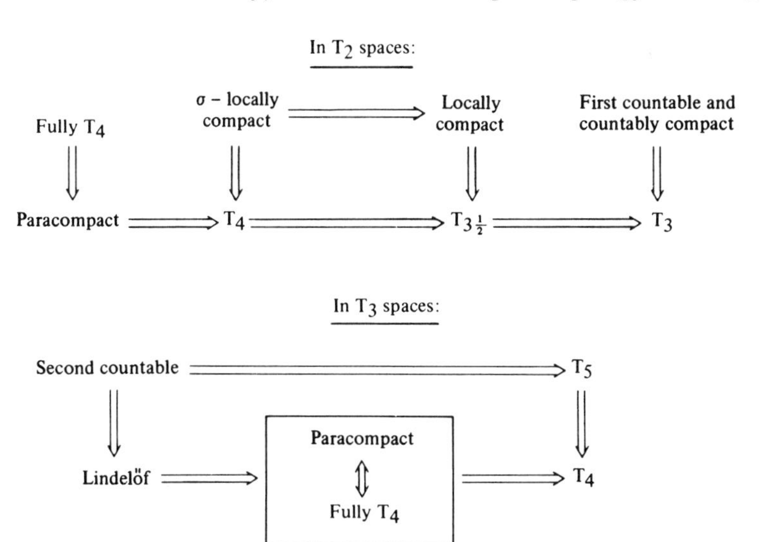

Figure 7.

the identity map $f\colon (X,\tau^*) \to (X,\tau)$ is a continuous bijection of a Hausdorff space to a compact Hausdorff space. If τ^* is also compact, f must be open, hence $\tau^* \subset \tau$. Thus no topology strictly larger than τ can be compact. Examples 99 and 100 show that the converse statements are not necessarily true: minimal Hausdorff topologies need not be compact, and maximal compact topologies need not be Hausdorff.

We should note, finally that separable Hausdorff spaces can have cardinality not exceeding $2^{2^{\aleph_0}}$, for if D is a countable dense subset of X, the map $\Phi\colon X \to 2^{P(D)}$ defined by $\Phi(x)(A) = 1$ iff $A = D \cap U_x$ for some neighborhood U_x of x is one-to-one whenever X is Hausdorff. Thus card $(X) \leq \text{card } 2^{P(D)} = 2^{2^{\aleph_0}}$.

Invariance Properties

It can be easily seen that most global and local compactness properties (namely those defined by covers) are weakly hereditary, that is, they are preserved in closed subspaces. But in most cases they are not preserved in open subspaces, so are not hereditary. Paracompactness and metacompactness, similarly, are only weakly hereditary. Both first and second countability are, however, hereditary, although separability is not. In fact, separability is preserved only in open subspaces (Example 10.6).

Most compactness properties fail to be preserved by arbitrary products. The most famous exception is compactness itself, for, by Tychonoff's

Table 1

PROPERTIES PRESERVED BY PRODUCTS

Property of Factor Spaces	Type of Product: Finite	Countable	Uncountable
Compact	True	True	True
σ-compact	True	False (102)	False
Sequentially compact	True	True	False (105)
Countably compact	False (112)	False	False
Locally compact	True	False (102)	False
Lindelöf	False (84)	False	False
First countable	True	True	False (103)
Second countable	True	True	False (103)
Separable	True	True	False (103)
Paracompact	False (84)	False	False
Metacompact	False (84)	False	False

theorem, the product of an arbitrary family of topological spaces is compact iff each factor space is compact. If X is compact, then in general $X \times Y$ has the compactness properties of Y. If X satisfies only weaker conditions, the situation is considerably more complex, and may be best summarized by Table 1 which indicates which properties are preserved by various types of products, and cites counterexamples where appropriate.

Conversely, it is often possible to infer properties of the factors given a property of the product space. This may be done most easily by observing that the projection maps are continuous, but in general, only the global compactness properties are preserved under continuous mappings. To be precise, the properties of compactness, σ-compactness, countable compactness, sequential compactness, Lindelöf, and separability are preserved under continuous maps and therefore also under projections. Local compactness, and first and second countability are preserved under open continuous maps, but not just under continuous maps (Examples 116 and 26); since projections are open and continuous, these properties also are preserved under projection maps. Paracompactness even fails to be preserved under open continuous maps (Example 11.19), although it is preserved under projections.

SECTION 4
Connectedness

Connectedness denies the existence of certain subsets of a topological space with the property that $\bar{U} \cap V = \varnothing$ and $U \cap \bar{V} = \varnothing$. Any two such subsets are said to be **separated** in the space. Although this concept is logically related to the separation axioms, it examines the structure of topological spaces from the opposite point of view.

We call two open sets U and V a **separation** of a topological space X if $U \cap V = \varnothing$ and $X = U \cup V$; spaces which have no nontrivial separations are **connected**. Equivalently, X is connected iff it is not the union of two separated sets; or it is not the union of two disjoint, closed sets; or, it does not have any nontrivial sets which are both open and closed; or, there is no continuous function from X onto the two point set, with the discrete topology. A connected space X is said to be **degenerate** if it consists of a single point. A subset in a topological space X is a **connected set** if it is not the union of two separated subsets of X, or, equivalently, if it satisfies the definition of a connected space under the induced topology. Two points of X are **connected in X** if there exists a connected set containing them both. This relation between the points of a space is an equivalence relation, since the union of any family of connected sets having a nonempty intersection is connected. The disjoint equivalence classes of points of X under the relation "connected in X" are called the **components** of X. The components of X are precisely the maximal connected subsets of X, and they must be closed since the closure of every connected set is connected: any separation of $\bar{E}$ would either separate E, or separate E from some of its limit points. (This shows even more, namely, if $E \subset F \subset \bar{E}$ and if E is connected, then F is connected.) Each nonempty set in X which is both open and closed contains the components of all of its points, but the component of a point need not coincide with intersections of the sets containing it which are both open and closed (Example 115). We say that a space X is **connected between two points** if each separation of X includes a single open set containing both points. This too is an equivalence relation between the points of a space; we will call the equivalence classes the **quasicomponents**. The quasicomponent containing $p \in X$ is precisely the intersection of all sets containing p which are both open and closed. If a space X has just one quasicomponent, it must in fact be connected; thus we need not call it quasiconnected.

Path and arc connectedness relate to the existence of certain continuous functions from the unit interval into a topological space. Continuous functions from the unit interval are called **paths**; if they are one-to-one they are **arcs**. A space is **path connected** if for every pair of points a and b there exists a path f such that $f(0) = a$ and $f(1) = b$. The existence of a path between two points of a space is an equivalence relation; transitivity may be verified by reparametrizing the two paths. The equivalence classes, called **path components**, are the maximal subsets with respect to path connectedness. **Arc connectedness** and **arc components** are defined by exact analogy; to make the relation reflexive, we declare every point arc connected to itself. Clearly, every nontrivial arc connected space must be uncountable.

The relations between the four types of components may be summarized by the following chain of containments:

$$\begin{matrix}\text{Arc} \\ \text{components}\end{matrix} \subset \begin{matrix}\text{Path} \\ \text{components}\end{matrix} \subset \text{Components} \subset \text{Quasicomponents.}$$

None of these containments is reversible (Examples 8, 116, and 115).

A set with no disjoint open sets will be called **hyperconnected** and a set with no disjoint closed sets will be called **ultraconnected**. Equivalently, X is hyperconnected if the closure of every open set is the entire space, while X is ultraconnected if the closures of distinct points always intersect. Ultraconnectedness is independent of hyperconnectedness, though both imply connected. In fact, every ultraconnected space is path connected, for if p is a point in $\overline{\{a\}} \cap \overline{\{b\}}$, then the function $f\colon [0,1] \to X$ which maps each point of $[0,\frac{1}{2})$ to a, each point $(\frac{1}{2},1]$ to b, and $\frac{1}{2}$ to p is continuous. Hyperconnected spaces need not be path connected (Example 18) and ultraconnected spaces need not be arc connected (Example 13). So we may summarize the connectedness implications by:

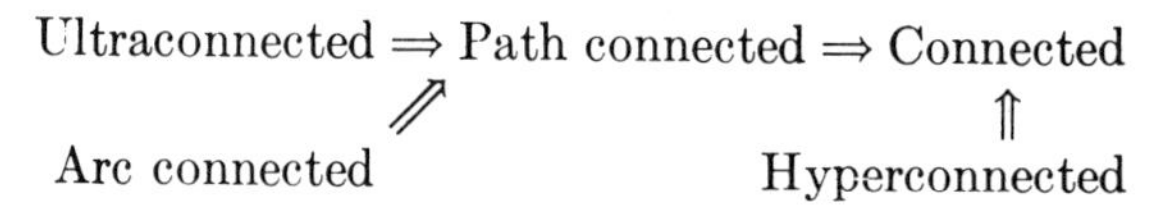

Both ultra- and hyperconnectedness are very strong conditions which trivially imply some other properties. Every continuous real-valued function on a hyperconnected space is constant, so such spaces are necessarily pseudocompact. On the other hand, no nontrivial ultraconnected space can have more than one closed point, so none are T_1, even though they must all be T_4, trivially.

Quasicomponents and components are equal if (but not only if; see Example 26) a space has a basis consisting of connected sets; we call such a space **locally connected**. Equivalently, X is locally connected if the components of open subsets of X are open in X. Local connectedness clearly does not imply connectedness, but neither does connectedness imply local connectedness (Example 116). However, every hyperconnected space is clearly locally connected, since in such spaces every open set is connected. Figure 8 summarizes the relevant counterexamples.

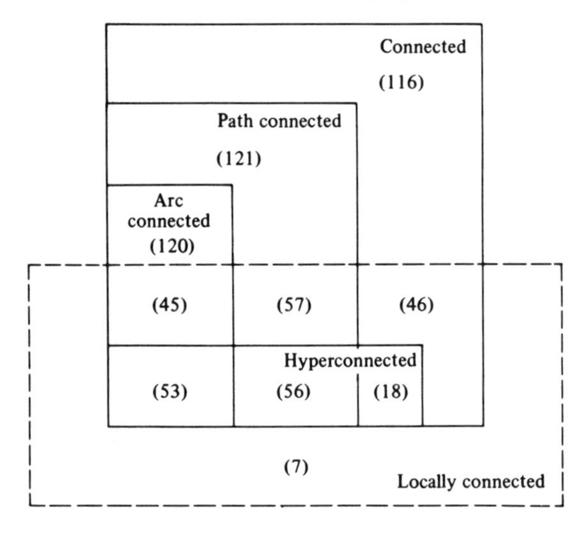

Figure 8.

Path components are equal to quasicomponents if a space has a basis consisting of path connected sets; such a space is called **locally path connected**. Equivalently, X is locally path connected if the path components of open subsets of X are open in X. Analogously, arc components are equal to quasicomponents if a space has a basis of arc connected sets; such a space is said to be **locally arc connected**. As above, locally arc connected implies locally path connected, which implies locally connected, but neither converse holds (Examples 4 and 18). Furthermore, locally path connected is independent of path connected and locally arc connected is independent of arc connected (Examples 118 and 32.5).

Functions and Products

Any set S which is the union of connected sets A_α and a connected set B where $B \cap A_\alpha \neq \emptyset$ for each α must be connected since a separation of S would necessarily separate B. Since any finite product $\prod_{i=1}^{n} X_i$ of connected sets X_i can be written as the union of spaces homeomorphic to $\prod_{i=1}^{n-1} X_i$ and X_n, a simple induction argument shows that any finite product of connected spaces is connected. In fact, a straightforward argument by transfinite induction can be used to show that any product $\prod_{\alpha \in A} X_\alpha$ of connected spaces X_α is connected. If the index set A is well ordered and if $x = \langle x_\alpha \rangle \in X = \prod X_\alpha$ is some fixed point, let $S_\alpha = \{\langle y_\beta \rangle \in X | y_\beta = x_\beta \text{ for all } \beta \geq \alpha\}$. Then S_α is connected whenever $S_{\alpha-1}$ is since S_α is homeomorphic to $S_{\alpha-1} \times X_\alpha$. If α is a limit ordinal, $S_\alpha = \overline{\bigcup_{\beta < \alpha} S_\beta}$, so if each S_β is connected for $\beta < \alpha$, S_α must be also, since the collection $\{S_\gamma\}$ is nested. Thus $X = \overline{\bigcup_{\alpha \in A} S_\alpha}$ is connected. Indeed we have proved more since the proof uses only the facts that in the product topology the subsets $X'_\alpha \subset \prod X_\alpha$ where $X'_\alpha = \{\langle y_\beta \rangle \in X | y_\beta = x_\beta,\ \beta \neq \alpha\}$ are homeomorphic to the X_α's and that $X = \overline{\bigcup_{\alpha \in A} S_\alpha}$. Thus this proof applies to the Cartesian product of the X_α with any topology in which the sets X'_α are copies of the corresponding X_α, and $X = \overline{\bigcup_{\alpha \in A} S_\alpha}$.

If X is connected and f is a continuous function on X, then $f(X)$ must be connected, for if A and B separate $f(X)$, $f^{-1}(A)$ and $f^{-1}(B)$ separate X. Though the continuous image of a locally connected space need not be locally connected, it is true that local connectedness is preserved under continuous maps f from a compact space X onto a Hausdorff space Y. For suppose E is a component of an open subset U of Y. Then each component of $f^{-1}(E)$ is a component of $f^{-1}(U)$ since if G is a component of $f^{-1}(U)$, then $f(G)$ is connected and thus either contained in E or disjoint from it. But if X is locally connected, the components of the open set $f^{-1}(U)$ are open, so $f^{-1}(E)$ must be open. Its complement is closed, thus compact, so $f(X - f^{-1}(E)) = Y - E$ is compact, hence closed (since Y is Hausdorff). Thus E is open, and therefore Y must be locally connected.

Disconnectedness

A space is **totally pathwise disconnected** if the only continuous maps from the unit interval into X are constant, or, equivalently, if its path components are single points. A space with single point components is said to be **totally disconnected**; since the points will then be closed, each such space will be T_1. Clearly no connected set with more than one point can be totally disconnected, though there is a connected set which is totally pathwise disconnected (Example 128). Furthermore, only the discrete space can be both totally disconnected and locally connected, for in such cases each component (that is, each point) must be open.

If for every pair of points a and b in a space X there exists a separation U, V such that $a \in U$ and $b \in V$, we shall say that X is **totally separated**. A necessary and sufficient condition that X be totally separated is that its quasicomponents be single points; clearly every totally separated space is completely Hausdorff and Urysohn. A Hausdorff space in which the closure of every open set is open is called **extremally disconnected**;

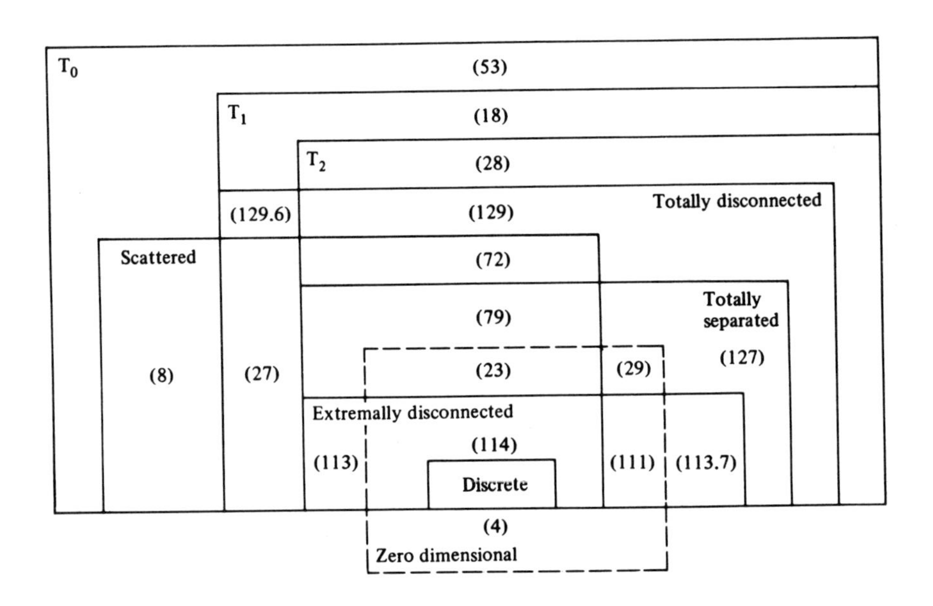

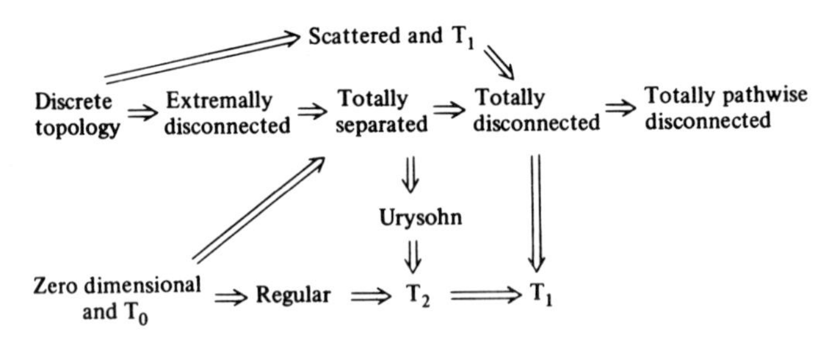

Figure 9.

equivalently, a Hausdorff space is extremally disconnected iff the interior of every closed set is closed, or, if disjoint open sets always have disjoint closures. Clearly every extremally disconnected space is totally separated.

A space is **zero dimensional** if it has a basis consisting of sets which are both open and closed. Clearly every zero dimensional space is T_3, though not necessarily T_0 (Example 5). Every zero dimensional T_0 space is totally separated. A space is called **scattered** if it contains no nonempty dense-in-itself subsets; although scattered spaces need not be T_1, every scattered T_1 space must be totally disconnected since in T_1 spaces, every nontrivial connected set is dense-in-itself. However, a scattered space which is not T_1 may be connected (Example 57). Thus we may summarize the various disconnectedness properties in Figure 9.

Biconnectedness and Continua

A connected set is said to be **biconnected** if it is not the union of two disjoint nondegenerate connected subsets. A point p of a connected set X is called a **cut point** if $X - \{p\}$ is disconnected, and a **dispersion point** if $X - \{p\}$ is totally disconnected; any set having a dispersion point is biconnected, since the dispersion point can be in at most one of the two disjoint subsets. There is, however, a biconnected set without a dispersion point (Example 131).

Sets which are both compact and connected are called **continua**; a continuum is **indecomposable** if it is not the union of two different nondegenerate proper sub-continua. A subset C of a continuum K is a **composant** if for some $p \in K$, C contains all points x such that x and p are contained in some proper sub-continua of K. A set is said to be **puncti-form** if it contains no nondegenerate continua. Clearly, each totally disconnected space is punctiform, although so are some connected spaces (Example 128).

SECTION 5
Metric Spaces

A **metric** for a set X is a mapping d of $X \times X$ into the nonnegative real numbers satisfying the following conditions for all $x,y,z \in X$:

M_1: $d(x,x) = 0$
M_2: $d(x,z) \leq d(x,y) + d(y,z)$
M_3: $d(x,y) = d(y,x)$
M_4: if $x \neq y$, $d(x,y) > 0$.

We call $d(x,y)$ the **distance** between x and y. If d satisfies only M_1, M_2, and M_4 it is called **quasimetric**, while if it satisfies M_1, M_2, and M_3 it is called a **pseudometric.** It is possible to use a metric to define a topology on X by taking as a basis all **open balls** $B(x,\epsilon) = \{y \in X | d(x,y) < \epsilon\}$. A topological space together with a metric giving its topology is called a **metric space.** Although a single metric will yield a unique topology on a given set, it is possible to find more than one metric which will yield the same topology. In fact, there are always an infinite number of metrics which will yield the same metric space (Example 134).

Every metric space is Hausdorff, since $B(p,\epsilon) \cap B(q,\epsilon) = \varnothing$, if $\epsilon < d(p,q)/2$, and also T_5. For suppose A and B are separated subsets of a metric space X; then each point $x \in A$ has a neighborhood $B(x,\epsilon_x)$ disjoint from B, and each point $y \in B$ has a neighborhood $B(y,\epsilon_y)$ disjoint from A. Then $U_A = \bigcup_{x \in A} B(x,\epsilon_x/2)$ and $U_B = \bigcup_{y \in B} B(y,\epsilon_y/2)$ are disjoint open neighborhoods of A and B, respectively. Thus metric spaces are completely normal, and, by a similar argument, perfectly normal. Therefore metric spaces satisfy every T_i separation property. Furthermore, every metric space is fully T_4, thus fully normal and paracompact.

Much of the structure of countability and compactness is also simplified in metric spaces. Since $\{B(x,1/n) | n = 1, 2, 3, \ldots\}$ is a countable local basis at x, each metric space is first countable. If $\{x_i\}$ is a countable dense subset of X, the balls $B(x_i,1/n)$ form a countable base for the topology on X. So for metric spaces separability implies, and is therefore equivalent to, second countability.

The same is true of metric spaces which are Lindelöf, since in such spaces, for each integer k, the open covers $\{B(x,1/k)|x \in X\}$ have countable subcovers. The union of all such subcovers is a countable base for X. Thus every Lindelöf metric space is also second countable.

Since each metric space X is first countable, sequential compactness is equivalent to countable compactness, which, since X is T_1, is equivalent to weak countable compactness. More important, countable compactness in metric spaces is equivalent to compactness, since every countably compact metric space is separable: for each n, a countably compact metric space can be covered by finitely many balls $B(x_i^n,1/n)$, so $\{x_i^n\}$ is a countable dense subset. Thus each countably compact metric space is second countable, and every countably compact second countable space is compact.

Since metric spaces are Hausdorff, the concepts of local compactness and strong local compactness are equivalent. So in metric space, we have a much simplified implication chart (Figure 10); that these implications do not reverse is shown by the counterexamples listed in Figure 11.

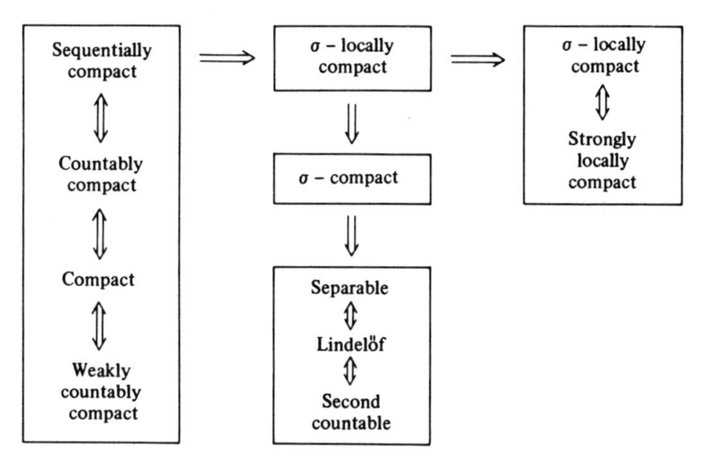

Figure 10.

Although in general the metric structure of a space does not appreciably simplify its connectedness properties, we can show that every metric space which is extremally disconnected is discrete. For in any metric space, each point p can be written as the intersection of the closed metric balls

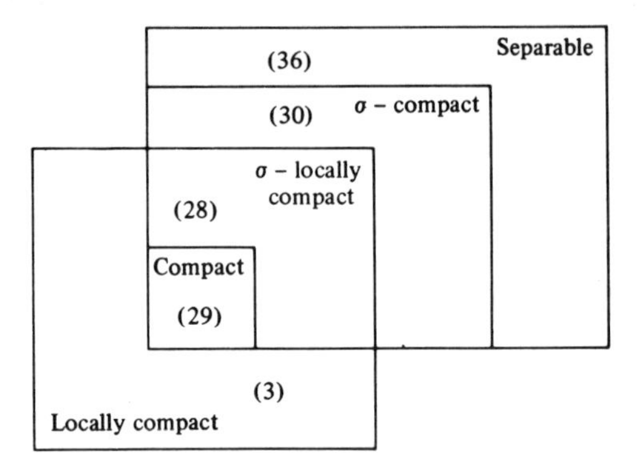

Figure 11.

$\overline{B_{1/n}(p)}$; then $U = \bigcup_{n=1}^{\infty} B_{1/2n}(p) - \overline{B_{1/(2n+1)}(p)}$ is an open set which has p as a noninterior limit point provided p is not open. So if X is not discrete, it cannot be extremally disconnected.

Complete Metric Spaces

Heuristically, compactness is related to the size of a space in that it determines how many small open sets are required for a cover. In a metric space, the radius ϵ of an open ball can be used as a precise measure of the size of the small open sets. Thus we call a subset E of a metric space X **totally bounded** (**precompact**) iff for every $\epsilon > 0$, E may be covered by a finite collection of open balls of radius ϵ. We call such a cover an **ϵ-net.** A subset E is called **bounded** if there exists a real number β such that $d(x,y) \leq \beta$ wherever $x,y \in E$; the least bound of E is called the **diameter** of E. Clearly every totally bounded set is bounded, but not conversely (Example 134); furthermore, totally bounded is not a topological property, since it is not preserved by homeomorphisms (Example 134).

Every compact metric space is totally bounded, since every covering by ϵ-balls has a finite subcover, and every totally bounded set is second countable, since the union of ϵ-nets for $\epsilon = 1, \frac{1}{2}, \frac{1}{3}, \ldots$ forms a countable basis. But neither of these implications reverses (Examples 30.10 and 134).

To discover the reason that totally bounded sets may fail to be compact, we must examine the convergent sequences. A sequence $\{x_n\}$ in a metric space (X,d) is called a **Cauchy sequence** iff for every $\epsilon > 0$ there exists an integer N such that $d(x_m,x_n) < \epsilon$ wherever $m,n > N$. Obviously, every convergent sequence is a Cauchy sequence; but the converse fails (Example 32.1). So we define a **complete** metric space as one in which every Cauchy sequence converges to some point in the space, or equivalently that the intersection of every nested sequence of closed balls with radii tending to zero is nonempty. (A sequence $\{E_n\}$ of sets is **nested** iff $E_n \supset E_{n+1}$ for all n.) If the radii do not tend to zero, this condition need not be implied by completeness (Example 135). Now every compact metric space is complete, and more important, every complete and totally bounded metric space is compact.

We will call a topological space (X,τ) **topologically complete** if there exists a metric d giving the topology τ such that (X,d) is a complete metric space. Topological completeness is a topological property which is weakly hereditary, though not hereditary (Example 30). Clearly every compact metric space is complete; though the converse is not true (Example 28), it is true that a metric space is compact iff it is complete in every equivalent metric. The famous Baire category theorem states that every topologically complete metric space is second category.

A completion of a metric space X is any complete metric space which contains a dense subset to which X is **isometric**, that is, to which there is a bijection which is distance preserving. All metric spaces have completions and even more surprising, all of the completions of a given space are isometric. Furthermore completeness is preserved by isometries but unlike topological completeness not by homeomorphisms (Example 32.10).

Metrizability

A topological space (X,τ) is called **metrizable** if there exists a metric d which yields the topology τ. Every regular second countable space is metrizable, but not conversely (Example 3); in fact, a topological space is metrizable iff it is regular and has a **σ-locally finite base**, that is, a base which is the countable union of locally finite families. Although this requirement is very close to paracompactness, and though every metric space is paracompact, there exist regular paracompact spaces which are nonmetrizable (Examples 51 and 141).

Uniformities

A **quasiuniformity** on a set X is a collection U of subsets of $X \times X$ which satisfies the following axioms:

U_1: For all $u \in U$, $\Delta \subset u$, where $\Delta = \{(x,x)|x \in X\}$.
U_2: For $u \in U$ and $v \in U$, $u \cap v \in U$.
U_3: If $u \in U$ and $u \subset v \subset X \times X$, then $v \in U$.
U_4: For all $u \in U$, there is $v \in U$ such that $v \circ v \subset u$ where $\circ$ is defined by $u \circ v = \{(x,z)|$ there is a $y \in X$ such that $(x,y) \in v$ and $(y,z) \in u\}$.

The quasiuniformity U is a **uniformity** if the following additional condition is satisfied:

U_5: If $u \in U$, then $u^{-1} \in U$ where $u^{-1} = \{(y,x)|(x,y) \in U\}$.

A set u which is an element of the quasiuniformity U is called an **entourage** (or a **relation**). The entourage u is said to be **symmetric** if $u = u^{-1}$. The set Δ is called the **diagonal** of $X \times X$. The quasiuniformity U is said to be **separated** if the intersection of all the members of U is the diagonal Δ.

The first three axioms say that every quasiuniformity on a set X is a fiiter on $X \times X$. Further, a quasiuniformity U is a uniformity iff there is a **symmetric base** for U, that is a filter base of symmetric sets.

Every quasiuniformity U on a set X yields a topology τ on X by taking as a neighborhood system for X the sets $u(x)$ where $u \in U$ and $u(x) = \{y|(x,y) \in u\}$; there may be more than one quasiuniformity generating a given topology (Example 44). If two quasiuniformities generate the same topology on the set X, they are said to be **compatible**. A set X with a quasiuniformity U and the topology τ generated by U is said to be a **quasiuniform space** and we may use the notation $((X, U),\tau)$ to denote this or the shorter notation (X,U) where τ is understood to be the topology generated on X by U. A topological space (X,τ) is said to be **quasiuniformizable** if there is a quasiuniformity U such that $((X,U),\tau)$ is a quasiuniform space.

The problem of when a topological space (X,τ) is quasiunformizable or uniformizable is simpler than the corresponding metrization problem. If (X,τ) is a topological space, the set $U = \{u_G|u_G = (G \times G) \cup ((X - G) \times (X - G))$ and $G \in \tau\}$ is a filter subbase for a quasiuniformity on X which generates τ, and thus every topological space is quasiuniformizable. A topological space (X,τ) is uniformizable iff it is a $T_{3\frac{1}{2}}$ space.

Metric Uniformities

If (X,d) is a pseudometric space, then the family U of all sets u which contain a set of the form $u_\epsilon = \{\langle x,y\rangle|d(x,y) < \epsilon\}$ is a uniformity on X, which yields the same topology as the pseudometric d. Such a uniformity is called **pseudometrizable** (or, if appropriate, **metrizable**). Not every uniformity which yields a metrizable topological space need be metrizable (Example 44).

PART II
Counterexamples

1. Finite Discrete Topology

2. Countable Discrete Topology

3. Uncountable Discrete Topology

On any set X we define the discrete topology by taking all subsets of X to be open. Any subset is then both open and closed. We distinguish three cases, the finite discrete topology, the countable discrete topology, and the uncountable discrete topology according to whether the set X is finite, countably infinite, or uncountable.

1. This topology is the finest topology for X, since any open set of any other topology is an open set in this topology.

2. Every point is an isolated point.

3. x is not a limit point of the sequence $x, x, x, \ldots$ considered as a set, although it is an adherent point of the set.

4. For any set $A \subset X$, $A = A^{\circ} = A^{-}$, and $A^{b} = \emptyset$.

5. Any function from a set X with the discrete topology is continuous.

6. The topology on a discrete space may be obtained from the discrete metric: $d(x,y) = 1$ if $x \neq y$, and $d(x,y) = 0$ if $x = y$. Thus every discrete space satisfies all separation properties.

7. Each discrete space is strongly locally compact since each point is a neighborhood of itself. Such spaces are clearly first countable

and, since the open cover by discrete points is locally finite and finer than all other open covers, discrete spaces are paracompact.

8. Countable discrete spaces are σ-compact, Lindelöf, second countable, and separable but uncountable discrete spaces are none of the above. Finite discrete spaces satisfy all compactness properties.

9. Since only the empty set is nowhere dense, every discrete space is of the second category. (In fact, a discrete space is a complete metric space.) Furthermore, no discrete space is dense-in-itself.

10. If X consists of more than one point, it is clearly not connected and thus neither path nor arc connected. But it is locally path connected, and thus locally connected.

11. The topology on the discrete space is generated by the discrete uniformity which consists of all subsets of $X \times X$ which contain the diagonal Δ. The diagonal Δ is a base for this uniformity.

4. Indiscrete Topology

For any set X, the indiscrete topology is the topology whose only elements are the empty set $\varnothing$ and X itself. (We assume X has at least two points.)

1. This topology is the coarsest one for X. It is comparable with any other topology for X.

2. No subset $A \neq X$ or $\varnothing$ is open, closed, F_σ, or G_δ.

3. Every subset is compact and sequentially compact.

4. Every point of X is a limit point for every subset of X, and every sequence converges to every point of X. If X is uncountable, every sequence has uncountably many limit points.

5. Every subset containing more than one point is dense-in-itself. The only nowhere dense subset is $\varnothing$, so X is of the second category.

6. For $A \neq X$, $A^\circ = A^{\circ -} = A^{\circ - \circ} = \varnothing$ and for $A \neq \varnothing$, $A^- = A^{-\circ} = A^{-\circ -} = X$. If $A \neq X$ or $\varnothing$, $A^b = X$, $A^{bb} = \varnothing$.

7. X is separable, since any subset is dense. Furthermore, X is second countable.

8. Every function to a space with the indiscrete topology is continuous.

9. The indiscrete space is path connected and thus connected, but is arc connected only if it is uncountable. It is both hyperconnected and ultraconnected.

10. Since the only open set containing any given point is X itself, the indiscrete space fails to be T_0. But it is T_3, T_4, and T_5 vacuously.

11. Clearly X is pseudometrizable, although not metrizable.

5. Partition Topology

6. Odd-Even Topology

7. Deleted Integer Topology

Each partition P of any set X into disjoint subsets, together with $\varnothing$, is a basis for a topology on X, known as a partition topology. A subset of X is then open if and only if it is the union of sets belonging to P.

1. The partition topology is characterized by the fact that every open set is also closed; each set in the partition P is a component of the space X. Thus X/P is discrete.

2. The trivial partitions yield the discrete or indiscrete topologies. In any other case X with a partition topology is not T_0 since some element of the partition contains two or more points neither of which can be separated from the other. Thus X is not $T_{2\frac{1}{2}}$, T_2, or T_1. However a subset of X is open iff it is a union of elements of the partition and thus its complement is also open; thus a set is open iff it is closed. Hence X is T_3, $T_{3\frac{1}{2}}$, T_4, and T_5.

3. An important example of a partition topology is the odd-even topology on the set X of positive integers, generated by the partition $P = \{\{2k - 1, 2k\}\}$. Clearly this space is second countable, thus first countable, separable, and Lindelöf. Since every nonempty subset of X has a limit point in this topology, X is weakly countably compact. But X is not countably compact, since P itself is a countable open covering of X which has no finite subcover.

4. If X is the set of positive integers with the odd-even topology, and if Z^+ is the same set with the discrete topology, then the mapping $f\colon X \to Z^+$ defined by $f(2k) = k$, $f(2k - 1) = k$ is continuous. But X is weakly countably compact, whereas Z^+ is not.

So weak countable compactness is not preserved under continuous maps.

5. A common variation of the odd-even topology is the deleted integer topology: in this case X is the union of the open intervals $(n - 1,n)$ for $n = 1, 2, 3, \ldots$, and the topology on X is generated by the partition $P = \{(n - 1,n)\}$. This example has most of the properties of the odd-even topology.

6. Every partition space is pseudometrizable since the pseudometric defined by letting $d(x,y) = 0$ iff x and y belong to the same set of the partition, and letting $d(x,y) = 1$ otherwise yields the partition topology.

7. If we double the points of the real numbers with the discrete topology, we obtain a partition topology with uncountably many disjoint open sets. This topology is weakly countably compact but not Lindelöf.

8. Finite Particular Point Topology

9. Countable Particular Point Topology

10. Uncountable Particular Point Topology

11. Sierpinski Space

12. Closed Extension Topology

On any set X, we can define the open sets of a topology to be $\varnothing$ and any subset of X that contains a particular point p. We distinguish three cases, finite, countable, and uncountable according to the size of X.

1. The only sequences $\{a_i\}$ which converge are those for which the a_i are equal for all but a finite number of indices. The only accumulation points for sequences are the points b_j that the a_i equal for infinitely many indices. So any countably infinite set containing p has a limit point, but never even an accumulation point when considered as a sequence in any ordering.

2. Every point except p in X is a limit point of p, so the closure of any open set other than $\varnothing$ is X. Closed sets other than X do not contain p, so the interior of any closed set other than X is $\varnothing$.

3. Let Y be a subset containing the particular point p. Then every point $q \neq p$ is a limit point of Y but not an ω-accumulation point.

4. Every particular point topology is T_0, but since there are no disjoint open sets, none of the higher separation axioms are satisfied unless X has only one point.

5. $A = \{p\}$ is compact, but $\bar{A} = X$ is not compact if X is infinite. In this case, X is locally compact but not strongly locally compact, since the closure of any set containing p is X. In fact, if X is uncountable, it is not even Lindelöf.

6. X is separable, since $\{p\}$ is a countable dense subset. But, if X is uncountable, $X - \{p\}$ is not separable.

7. If X is uncountable, it is first countable, but not second countable, since $X - \{p\}$ is discrete.

8. If on a given set X, we define τ_1 to be the collection of all sets containing a point p, and τ_2 to be the collection of all sets containing $q \neq p$, the spaces (X,τ_1) and (X,τ_2) are homeomorphic, but τ_1 and τ_2 are not comparable.

9. X is scattered, since every subset not containing p has no limit point, and for a subset which contains p, p itself is not a limit point. Thus X contains no nonempty dense-in-itself subsets.

10. X is hyperconnected, since every open set must contain p. But if X contains at least three points, it is not ultraconnected since two points not equal to p are disjoint closed sets.

11. Since $X - \{p\}$ is discrete, p is a dispersion point for X.

12. X is not weakly countably compact since any set which does not contain p has no limit points. But since there are no disjoint open sets, every continuous real valued function on X is constant. Thus X is pseudocompact.

13. X is path connected and locally path connected since if $q \in X$ we can map 1 to q and $[0,1)$ to p to form a path from q to p. But X is not arc connected since the inverse image (under a homeomorphism) of the open set p would be one point, which is not an open set in $[0,1]$.

14. X is not of the first category, since if it were, some nowhere dense set would have to contain p, and its closure would then be X.

15. (X,τ) is locally compact since each point has a compact neighbor-

hood, namely itself together with p; but if X is infinite, it is not strongly locally compact since the closure of any neighborhood is all of X.

16. If we replace the particular point p with two points p_1, p_2, the resulting space is weakly countably compact since either p_1 or p_2 is a limit point of any subset.

17. An important particular point topology is Sierpinski space, the space $\{0,1\}$ with the particular point 0. Since the only open sets are $\emptyset$, X, and $\{0\}$, the sequence 0, 1, 0, 1, . . . has 0 as an accumulation point and 1 as a limit point.

18. Sierpinski space is hyperconnected, ultraconnected, and path connected, but not arc connected. Also it is T_4 and T_5 vacuously.

19. Let (X,τ) be a countable set with the discrete topology, which is then paracompact, and let $Y = \{0,1\}$ be Sierpinski space with 0 open, which is compact; then $X \times Y$ is paracompact. If $(X \cup \{p\},\sigma)$ is a particular point space with particular point p then the cover $\{\{p,\alpha\} | \alpha \in X\}$ is a countable cover with no point finite refinement. Thus $(X \cup \{p\},\sigma)$ is not even countably metacompact. However the function f: $X \times Y \to X \cup \{p\}$ defined by $f(x,0) = p$ and $f(x,1) = x$ is open and continuous. Thus the open continuous image of a paracompact space need not even be countably metacompact.

20. The particular point topology permits the following useful extension. Let (X,τ) be any nonempty space, and let p be a point not in X. We define $X^* = X \cup \{p\}$ and describe a topology τ^* on X^* by calling a set in X^* open iff it is the empty set or is of the form $U \cup \{p\}$ where $U \in \tau$. Since the closed sets of X^* other than X^* itself are precisely the closed sets of X we call (X^*,τ^*) the closed extension of (X,τ). The particular point topology on X is the closed extension of the discrete topology on $X - \{p\}$.

21. The properties of the closed extension topology are the same as the properties of the particular point topology except in the cases where the properties of the particular point topology depend on the discreteness of $X - \{p\}$. Thus (X^*,τ^*) is T_0 iff (X,τ) is T_0; but (X^*,τ^*) is not T_1, T_2, or T_3. Further (X^*,τ^*) is T_4 or T_5 iff (X,τ) is T_4 or T_5 vacuously and in this case the condition on (X^*,τ^*) is also vacuous.

13. Finite Excluded Point Topology

14. Countable Excluded Point Topology

15. Uncountable Excluded Point Topology

16. Open Extension Topology

The excluded point topology may be defined on any set X by declaring open, in addition to X itself, all sets which do not include a given point $p \in X$. As usual, we distinguish three special cases depending on the cardinality of X: finite, countable, or uncountable excluded point topology.

1. If X has just two points, the excluded point topology on X is just the Sierpinski topology. We consider this to be the trivial case, and assume hereafter that X has at least three distinct points.

2. X is T_0, but since the only neighborhood of p is X itself, X is not T_1, and thus not T_2 or T_3. However, every nonempty closed set contains p so X is T_4 vacuously. Since any two sets in X are separated iff they are disjoint subsets of $X - \{p\}$, and since such sets are open, X is T_5 nonvacuously.

3. Again, since X is the only open set containing p, X must be both compact and connected. Since every closed set other than $\varnothing$ contains p, X is ultraconnected, but it is not hyperconnected, since two points distinct from p are disjoint open sets. Thus X is path connected, though it cannot be arc connected since the inverse image of a single point distinct from p must be an open set in $[0,1]$. Similarly X is locally path connected but not locally arc connected.

4. Since $\{p\}$ is closed, and since the only open set which contains p is X itself, X is not perfectly T_4.

5. X contains no nonempty dense-in-itself subsets since only p can be a limit point of any set. Thus X is scattered. Further p is a dispersion point of X.

6. X is always first countable, and thus (since X is compact) sequentially compact. But it is second countable and separable only when X is finite or countable.

7. The excluded point topology may be varied by selecting as open

all sets which are disjoint from a fixed subset A, together with X itself. This excluded set topology is similar to the excluded point topology except that it will in general fail to be T_0.

8. The excluded point topology is a special case of the following. Let (X,τ) be a nonempty topological space, and let p be a point not in X. We define $X^* = X \cup \{p\}$ and describe a topology τ^* on X^* by calling a set in X^* open iff it is X^* or in τ. We call (X^*,τ^*) the open extension of (X,τ) since other than X^* itself the open sets of τ^* are just the open sets of τ. The excluded point topology is then the open extension of the discrete topology.

9. Except where the preceding arguments depend on the discreteness of $X - \{p\}$ the properties of (X^*,τ^*) are the same as those of the excluded point topology. Thus (X^*,τ^*) is T_0 iff (X,τ) is T_0 but (X^*,τ^*) always fails to be T_1, T_2, T_3. It is always T_4, but it is T_5 iff (X,τ) is. Similarly (X^*,τ^*) is compact, connected, and ultraconnected. Likewise (X^*,τ^*) is separable, first or second countable iff (X,τ) is.

10. The open extension of the particular point topology is T_0 and T_4, but neither T_1 nor T_5 since the particular point topology is only T_0.

17. Either-Or Topology

The either-or topology is defined on the interval $X = [-1,1]$ by declaring a set open iff it either does not contain $\{0\}$ or does contain $(-1,1)$. Thus $\{1\}$, $\{-1\}$, $\{-1,1\}$ and any set containing $\{0\}$ are the nontrivial closed sets.

1. A straightforward consideration of cases shows that X is T_0 and T_4, but neither T_1 nor T_3. In fact, X is T_5, since if A and B are separated sets neither of which contains 0, they are then open. But if one, say A, contains 0, then 0 cannot be in $\bar{B}$. So $\bar{B}$ can be only $\{1\}$, $\{-1\}$, or $\{-1,1\}$, and in any of these cases B and $X - B$ are disjoint open sets containing B and A.

2. Since any open cover of X must include an open set containing 0, X is compact, thus Lindelöf. But the subspace $X - \{0\}$ is discrete, thus not Lindelöf.

3. X is clearly first countable, although not separable since X contains uncountably many open points. X is not of the first category, since no open point can be contained in any nowhere dense set.

4. X is locally path connected since every point except 0 is open, and the neighborhood $(-1,1)$ of 0 is path connected: if $p \in (-1,1)$, the function which takes 0 to 0 and $(0,1]$ to p is a path joining 0 to p. Thus X is also locally connected, but not locally arc connected.

5. However X is scattered, since there are no nonempty dense-in-itself subsets, for 0 is the only possible limit point of any subset.

18. Finite Complement Topology on a Countable Space

19. Finite Complement Topology on an Uncountable Space

We define the topology τ of finite complements (or cofinite topology) on any set X by declaring open those sets with finite complements, together with $\varnothing$ (and X). Then the only closed sets are X, $\varnothing$, and finite sets. If X is finite, the topology of finite complements is the discrete topology. So, to avoid trivialities, we will assume that X is infinite, and distinguish two cases, the topology of finite complements on a countable space, and the topology of finite complements on an uncountable space.

1. Each point of X is a limit point of any infinite subset A, since then any open set of X contains a point of A. In particular, if A is countably infinite, $\bar{A} = X$, so X is separable.

2. The space X and every subspace of X is compact. If we have a collection of open sets covering X, any one of the sets will cover all but a finite number of points of X, say n points of X. We can choose n other sets of the collection, one for each point, and together these $n + 1$ open sets will constitute a finite subcover of X.

3. If X is uncountable, open sets are uncountable, so are not F_σ-sets. By complementation, closed sets are not G_δ-sets. Thus X is not perfectly T_4. In this case, the countably infinite sets are F_σ-sets, which are neither open nor closed, and the complements of countably infinite sets are G_δ-sets, also neither open nor closed.

4. For uncountable X, this topology is not first countable, and therefore not second countable. Suppose at some point x there exists a countable local basis. Then there exists a countable collection of open sets $\mathcal{B}_x$, each containing x, such that every open neighborhood of x contains some set $B \in \mathcal{B}_x$. So $\bigcap \mathcal{B}_x = \{x\}$, and thus $X - \{x\} = X - \bigcap \mathcal{B}_x = \bigcup_{B \in \mathcal{B}_x} (X - B)$. Each of

the $X - B$ are finite by definition, and the countable union of finite sets is countable, so $X - \{x\}$ must be countable, a contradiction. But X is separable, since any infinite set is dense.

5. If $a,b \in X$, then $O_a = X - \{b\}$ is an open set containing a but not b, and $O_b = X - \{a\}$ contains b but not a, so X is a T_1 space.
6. Since no two open sets are disjoint (since X is assumed infinite), (X,τ) is hyperconnected and therefore not T_2, T_3, T_4, or T_5.
7. If one doubles the points of X, the resulting space satisfies no T_i axioms, but is still compact.
8. τ is the smallest (or coarsest) topology on X in which points are closed, thus it is often called the minimal T_1 topology.
9. Since X is hyperconnected it is connected and locally connected.
10. If X is countable, it cannot be path connected for if f: $[0,1] \rightarrow X$ is continuous, $\{f^{-1}(x)|x \in X\}$ is a countable collection of mutually disjoint closed sets whose union is $[0,1]$. But this is impossible. For the same reason, X cannot be locally path connected.
11. If X is uncountable, and if we assume the continuum hypothesis, then each pair of points a, $b \in X$ is contained in some set S whose cardinality is that of $[0,1]$. If f: $[0,1] \rightarrow S$ is a bijection, it is continuous, so S is an arc in X joining a and b. Thus in this case X is arc connected and similarly, locally arc connected.

20. Countable Complement Topology

21. Double Pointed Countable Complement Topology

If X is an uncountable set, we define the topology of countable complements on X by declaring open all sets whose complements are countable, together with $\varnothing$ (and X).

1. Since the topology of countable complements is finer than the minimal T_1 topology, it is T_1 and T_0; but it does not satisfy any other T_i axioms since no two open sets are disjoint.
2. Since the only compact sets are finite sets the space is neither σ-compact nor countably compact, though since the complement of any open set is countable the space is Lindelöf.
3. X is not even first countable for the same reason that the topology of finite complements is not. Since no countable set has a limit

point, X is not separable even though it satisfies the countable chain condition.

4. X is hyperconnected and thus connected, locally connected, and pseudocompact.

5. Since in this topology the intersection of any countable collection of open sets is open and thus uncountable, X is not even countably metacompact.

6. An interesting variation of this space may be constructed by doubling each of its points. Technically, this double pointed countable complement topology is the product of X, the topology of countable complements, with the two point indiscrete space. Clearly the double pointed countable complement topology fails to be T_0 or T_1 and since each doublet is closed and no two open sets are disjoint, it fails to satisfy any higher T_i axioms.

7. The double pointed countable complement topology is weakly countably compact, since if p belongs to any infinite set A, then its twin p' is a limit point of A. (The ordinary topology of countable complements is not weakly countably compact.)

8. If we further vary this example by forming the open extension of the double pointed countable complement topology, we will have a space which is T_4 (since all open extension topologies are T_4) but not T_0, T_1, T_2, T_3, or T_5.

22. Compact Complement Topology

On (R,τ) the Euclidean space of real numbers, we define a new topology by letting $\tau^* = \{X \subset R | X = \emptyset \text{ or } R - X \text{ is compact in } (R,\tau)\}$. Since the compact sets in (R,τ) are closed under arbitrary intersection and finite unions, τ^* is a topology.

1. Since finite sets are compact in (R,τ), the topology τ^* is finer than the topology of finite complements. Thus (R,τ^*) is T_1.

2. However, no two open sets in (R,τ^*) can be disjoint, for the complement of their intersection, being the union of their compact complements, cannot be R. Thus X must fail to be T_2, and thus cannot be $T_{2\frac{1}{2}}$, nor, since it is T_1, can (R,τ^*) be T_3, T_4, or T_5.

3. For precisely the same reason, (R,τ^*) is hyperconnected, thus connected and locally connected. But it is not ultraconnected.

4. (R,τ^*) is compact, since if $\{O_\alpha\}$ is an open covering of R, $R - O_{\alpha_0}$

is compact in the Euclidean topology (for any $O_{\alpha_0} \in \{O_\alpha\}$). Since each O_α is open in the Euclidean topology, a finite number of them must cover $R - O_{\alpha_0}$.

5. Sets of the form $(-\infty,n) \cup (p - 1/n,\ p + 1/n) \cup (n,\infty)$ form a countable local basis at $p \in R$. So (R,τ^*) is first countable and similarly, also second countable. Thus it is also sequentially compact.

6. Since (R,τ^*) is coarser than the Euclidean topology, the rationals remain dense in the new topology. Thus (R,τ^*) is separable.

23. Countable Fort Space

24. Uncountable Fort Space

If X is any infinite set, and p a particular point of X, we can define a topology on X by declaring open any set whose complement either is finite or includes p. If X is countably infinite, we call this space countable Fort space; if X is uncountable, then uncountable Fort space.

1. This topology is clearly the minimal topology generated by the excluded point topology together with the topology of finite complements, and thus (X,τ) is T_1.

2. (X,τ) is T_5, since if A and B are separated sets and neither contains p, they are both open. Otherwise, since A and B are disjoint, p is in exactly one of them, say A; then B is open, but so too is $X - B$. For if B were not closed it would contain infinitely many points, while every open set containing p has a finite complement. So p, which belongs to A, would be in $\bar{B}$, and thus $A \cap \bar{B}$ could not be empty. So (X,τ) is completely normal, and thus satisfies all weaker separation conditions.

3. If X is uncountable, $\{p\}$ is a closed set which is not a G_δ-set, for every countable intersection of open sets containing p contains all but a countable number of points of X. Thus in this case, (X,τ) is not perfectly normal—that is, not all of its closed sets are G_δ-sets. In the countable case, X is perfectly normal.

4. X is compact, since any open covering of X must contain a neighborhood U of p whose complement is finite. So U, together with one neighborhood containing each point of $X - U$, is a finite subcover. Furthermore, X is sequentially compact since every sequence of infinitely many distinct points contains a subsequence

which converges to p. In fact, if X is uncountable, p is the only limit point of any infinite countable subset, so such an X cannot be separable.

5. If X is uncountable, it is not first countable either, since p cannot have a countable local base. For suppose $\{U_i\}$ is a countable collection of neighborhoods of p; then $X - U_i$ is finite, so $\bigcup_{i=1}^{\infty} (X - U_i) = X - \bigcap_{i=1}^{\infty} U_i$ is at most countable. Thus $\bigcap_{i=1}^{\infty} U_i \neq p$, so there exists a point $q \neq p$ in $\bigcap U_i$, and hence $X - q$ is a neighborhood of p which does not contain any U_i. So $\{U_i\}$ is not a neighborhood base of p.

6. But if X is countable, it must be separable (it is a countable dense subset of itself) as well as second countable, for the total number of neighborhoods of p—the only point in question—is countable, being in one-to-one correspondence with the totality of all finite sets. So countable Fort space, since it is regular, must be metrizable.

7. Every point q of X, except p, is both open and closed, so $\{q\}$ and $X - \{q\}$ separate X. Thus X is totally separated. But X is not extremally disconnected, for if A is an infinite set with an infinite complement which contains p, then A is open, although $\bar{A} = A \cup \{p\}$ is not open.

8. X is scattered since in T_1 spaces, every dense-in-itself subset must contain an infinite number of points. But this is impossible in X, since every point except p is open.

9. Since every set containing p is closed, p has a local basis of open and closed sets. Since each other point is open, X will be zero dimensional.

25. Fortissimo Space

If X is any uncountable set, and p a particular point of X, we can define a topology on X by declaring open any set whose complement either is countable or includes p.

1. This space, like uncountable Fort space, is completely normal and, like the countable complement topology, is Lindelöf but not compact, separable, or first countable, and thus not metrizable. But it is paracompact, since every open cover has a refinement con-

sisting of one special open set which contains p together with open points.

2. From observation of an infinite complement of a neighborhood of the point p it follows that X is not sequentially compact or even weakly countably compact. In fact only finite sets are compact so X is not σ-compact.

3. X is not pseudocompact since the function which maps a neighborhood of p to 0 and the elements of its countable complement one-to-one onto the remaining integers in R is continuous.

4. If we double the points of X we obtain a space that is weakly countably compact and Lindelöf but still neither σ-compact nor pseudocompact.

26. Arens-Fort Space

Let (X,τ) be the set of all ordered pairs of nonnegative integers with each pair open except (0,0). Open neighborhoods U of (0,0) are defined so that for all but a finite number of integers m, the sets $S_m = \{n|(m,n) \not\in U\}$ are each finite. Thus each open neighborhood of the origin contains all but a finite number of points in each of all but a finite number of columns.

1. τ contains Fort's topology with particular point (0,0), so it is T_2, T_1, T_0.

2. τ is T_5 for the same reason that Fort space is. The only nontrivial case occurs when A and B are separated sets where $(0,0) \in A$. Then B is open. But every point other than (0,0) is an interior point of the complement of B, so B is closed iff $(0,0) \notin \bar{B}$ and if $(0,0) \in \bar{B}$ then $A \cap \bar{B} \neq \emptyset$ so A and B are not separated. Thus (X,τ) is completely normal.

3. (X,τ) is not first countable for no sequence $\{x_n\}$ in $X - \{(0,0)\}$ can converge to (0,0). If $\{x_n\}$ contains points in at most finitely many columns, it clearly cannot converge to (0,0). If not, then we can find an infinite subsequence $\{y_n\}$ with at most one point in each column. But then $X - \{y_n\}$ is a neighborhood of (0,0), so neither $\{y_n\}$ nor $\{x_n\}$ could converge to (0,0).

4. Since X is countable, it is separable, σ-compact, and Lindelöf. But no neighborhood of (0,0) is compact and hence X is not locally compact, and thus neither compact nor countably compact.

5. Since X is T_3 and Lindelöf, it is paracompact. This can also be shown directly by selecting from an arbitrary open covering of X one set, say U, which contains (0,0). Then the covering consisting of U together with all of the open points in $X - U$ is a locally finite refinement of the original cover.

6. The identity mapping of X with the discrete topology onto (X,τ) is a continuous function from a space which is both first and second countable to one which is neither.

7. (X,τ) is neither connected nor locally connected since every neighborhood U of (0,0) has a separation, namely, $U - \{p\}$ and $\{p\}$, where $p \in U$, $p \neq (0,0)$.

8. Every neighborhood of (0,0) is closed since its complement consists of a discrete set of points, so X is zero dimensional, and thus totally separated. Also since every point but (0,0) is isolated, X is scattered, and not first category.

9. X is not extremally disconnected, for the closure of the set $S = \{(m,n)|m \text{ is even}\}$ is $S \cup \{(0,0)\}$. But this set is not open, for it does not contain any neighborhood of (0,0).

27. Modified Fort Space

Let the set X be the union of any infinite set N and two distinct one point sets $\{x_1\}$ and $\{x_2\}$. We topologize X by calling any subset of N open and calling any set containing x_1 or x_2 open iff it contains all but a finite number of points in N.

1. X is compact for in any cover there is some open set containing x_1, the complement of this set is then finite and hence covered by a finite subcover.

2. X is T_1 for each point in N is open and both x_1 and x_2 have neighborhoods not containing any other given point.

3. X is not T_2 for x_1 and x_2 do not have any disjoint neighborhoods, thus X is not T_3, T_4, or T_5.

4. Every point of X is a component since every set containing more than one point is separated, thus X is totally disconnected, and not locally connected.

5. If $X = A \cup B$ is a separation of X and $x_1 \in A$ then A is a closed and open set containing x_1. Then since the closure of any open set containing x_1 contains x_2, $x_2 \in A$. Thus the quasicomponent of x_1

contains $\{x_1\} \cup \{x_2\}$. But no point of N is in the quasicomponent of $\{x_1\}$. Thus $\{x_1\} \cup \{x_2\}$ is a quasicomponent though every component consists of a single point. So X is totally disconnected but not totally separated, and not zero dimensional.

6. X is scattered since it is T_1, and thus any dense-in-itself subset must be infinite; but any set with three or more points contains an isolated point.

7. X is sequentially compact since any sequence has either one point repeated infinitely many times or infinitely many distinct points In the first case the subsequence of repeated points converges to itself, while in the second case the subsequence of distinct points converges to both x_1 and x_2.

28. Euclidean Topology

We define the Euclidean (or, usual) topology on the set R of real numbers by using as a basis sets of the form $(a,b) = \{x \in R | a < x < b\}$.

1. The Euclidean topology on R is generated by the metric $d(x,y) = |x - y|$, where $|x|$ denotes, as usual, the absolute value of the real number x. So the metric space R satisfies all of the separation axioms. Furthermore, R is complete, so of the second category.

2. R is second countable (and therefore first countable and Lindelöf) since sets of the form (a,b) where a and b are rational, form a countable basis for R. Since the rationals are a countable dense subset of the reals, R is separable.

3. If $\{a_n\}$ is the sequence 1, 1, 1, 2, 1, 3, . . . , 1 is the only accumulation point of the sequence, but is not a limit point of the sequence.

4. R is not countably compact, since the open intervals $(n,n + 2)$, $n = 0, \pm 1, \pm 2, \ldots$ cover R but no finite subcollection covers R. But R is locally compact and σ-compact, since the closed and bounded intervals $[a,b]$ are compact.

5. Every closed subset A of R is a G_δ-set since $A = \bigcap_{n=1}^{\infty} A_n$ where A_n is a neighborhood of A of radius $1/n$—that is, $A_n = \bigcup_{x \in A} B(x,1/n)$. Each point not in A is contained in an ϵ-ball which is disjoint from A, and thus disjoint from some A_n.

6. Any open cover for R covers each of the compact intervals $[n,n+1]$, so an open cover can be reduced to a sequence of finite subcovers $\{G_i^{(n)}\}$ for each interval $[n,n+1]$. Then the sets $G_i^{(n)} \cap (n-1,n+2)$ form a locally finite refinement of the original open cover. Thus R is paracompact.

7. The topology on R can be given also by a quasimetric such as $d(x,y) = y - x$ if $y \geq x$, and $d(x,y) = 2(x-y)$ if $y < x$. Basis neighborhoods are off-center intervals, since points to the right of x are closer to it than are points to the left.

8. The collection of sets $S_{ab} = \{(x,y) | x,y < \mathrm{b} \text{ or } x,y > a\}$, where $a, b \in R$ and $a < b$, is a subbase for a uniformity U which generates the usual topology on R, but U is clearly not the usual metric uniformity.

9. Euclidean n-space R^n is defined to be the product of n copies of R. The product topology is that generated by the basis of open rectangles, sets formed by the cross product of one open interval from each copy of R. An equivalent basis consists of open n-spheres, the metric balls under the metric $d(x,y) = [\Sigma(x_i - y_i)^2]^{1/2}$.

29. The Cantor Set

The Cantor set C consists of all points in the closed unit interval which can be expressed to the base 3 without using the digit 1. This representation of points of C is unique, for even though many rational numbers have two possible ternary expansions—such as $\frac{1}{3} = 0.10000\ldots = 0.022222\ldots$—no number can be written in more than one way without using the digit 1.

1. Geometrically, the Cantor set is the set obtained by deleting a sequence of open sets, known as middle thirds, from the closed unit interval. The exact construction is as follows. From the closed interval $E_1 = [0,1]$, first remove the open interval $(\frac{1}{3},\frac{2}{3})$, leaving $E_2 = [0,\frac{1}{3}] \cup [\frac{2}{3},1]$. From the remaining intervals, delete the open intervals $(1/9, 2/9)$ and $(7/9, 8/9)$. Four closed intervals will remain; E_3 will denote their union. From these four, remove middle thirds as before, leaving E_4, the union of eight closed intervals. The Cantor set C is then the intersection of the successive closed remainders: $C = \bigcap_{i=1}^{\infty} E_i$.

2. The Cantor set is closed and compact because it is the intersection of closed subsets of the unit interval which is compact. Thus C

is a complete metric space and therefore satisfies all T_i axioms. Furthermore, C is second countable since the unit interval is.

3. C is dense-in-itself since every open set containing a point $p \in C$ contains points of C distinct from p. Thus C is not scattered, and, since it is closed, it is perfect.

4. The Cantor set is nowhere dense in $[0,1]$ since it is closed, and no open interval in $[0,1]$ is disjoint from all the deleted open intervals of $[0,1]$. Being nowhere dense in $[0,1]$, C is obviously of the first category in the closed unit interval. But, being itself a complete metric space, it is of second category in itself.

5. The Cantor set is uncountable. We can define a function f from the Cantor set onto the uncountable set $[0,1]$ as follows. If $x \in C$ is written uniquely to the base 3 without using the digit 1, $f(x)$ is the point in $[0,1]$ whose binary expansion is obtained by replacing each digit "2" in the ternary expansion of x by the digit "1." Clearly all points in $[0,1]$ may be obtained by such a process.

6. The components of C are single points, for if $a < b$ are two points in C there exists a real number $r \notin C$ such that $a < r < b$. Then $A = C \cap [0,r)$ and $B = C \cap (r,1]$ is a separation of C where $a \in A$ and $b \in B$. Thus C is totally separated.

7. But C is not extremally disconnected, since $C \cap [0,1/4)$ and $C \cap (1/4,1]$ are disjoint open subsets of C with intersecting closures, since $1/4 = 0.02020202 \ldots$ belongs to both closures-

8. The countably infinite product $A = \prod_{n=1}^{\infty} A_n$, of the two point discrete space $A_n = \{0,2\}$ (for all n) is homeomorphic to the Cantor set. In C, basis elements consist of all sets of the form $\{y \mid |x - y| < \epsilon\}$ for $x \in C$ and ϵ a positive number. In ΠA_n, the sets of the form $\{\langle a_i \rangle \in \Pi A_n | a_i \text{ is fixed for } 1 \leq i \leq n\}$ form a basis for the product topology. The function f taking each point $\langle a_1, a_2, a_3, \ldots \rangle$ of ΠA_n to the point $0.a_1a_2a_3 \ldots$ is a homeomorphism of ΠA_n onto C. Clearly both f and f^{-1} are continuous, since they take basis elements to basis elements.

9. Since C is totally separated, it is not locally connected. But C is the countable product of copies of the locally connected discrete space $\{0,2\}$.

30. The Rational Numbers

31. The Irrational Numbers

Let Q be the set of rational numbers, $Q \subset R$. Then $R - Q$ is the set of irrationals. In each case we impose the topology induced by the usual topology on R.

1. If B_α is the set containing the single rational α, then $\bigcup_\alpha \bar{B}_\alpha = Q$ but $\overline{\bigcup_\alpha B_\alpha} = R$. Also $(\bigcup B_\alpha)^e = \varnothing$ while $\bigcap B_\alpha{}^e = R - Q$.

2. Q is an F_σ-set in R that is neither closed nor G_δ, since its complement is neither open nor F_σ. Thus $R - Q$ is a G_δ-set, in fact $R - Q = \bigcap_{\alpha \in Q} (R - \{\alpha\})$.

3. $\overline{Q \cap (R - Q)} = \varnothing$ but $\bar{Q} \cap \overline{(R - Q)} = R$.

4. The Euclidean metric makes both Q and $R - Q$ into metric spaces and thus they are completely normal and paracompact.

5. If $\{r_i\}$ is an enumeration of Q, we can define a new metric on R by

$$d(x,y) = |x - y| + \sum_{i=1}^{\infty} 2^{-i} \inf\left(1, \left| \max_{j \le i} \frac{1}{|x - r_j|} - \max_{j \le i} \frac{1}{|y - r_j|} \right| \right).$$

   The metric d adds to the Euclidean distance between x and y a contribution which measures the relative distances of x and y from the rationals Q. If $B_\epsilon(p)$ is a Euclidean metric ball and $\Delta_\epsilon(p)$ a d-metric ball, it is clear that $\Delta_\epsilon(p) \subset B_\epsilon(p)$. The converse fails since if r is rational and ϵ sufficiently small, $\Delta_\epsilon(r) = \{r\}$; hence in the metric space (R,d) the rationals are open.

   But if we restrict d to the irrationals $R - Q$, we can always find, for each ϵ, a δ so that $B_\delta(p) \subset \Delta_\epsilon(p)$. Thus the metric space $(R - Q,d)$ is homeomorphic to the Euclidean irrationals.

   But $(R - Q,d)$ is complete, since no sequence $\{x_n\}$ which converges in the Euclidean topology to a rational r_k can be Cauchy: for each x_n in such a sequence, there exists a term x_m (where $m > n$) such that $d(x_n,x_m) \geq |x_n - x_m| + 2^{-k}$. Of course, those sequences which are Cauchy converge to irrationals, so $R - Q$ is topologically complete.

6. The complete metric space $R - Q$ is of the second category, while Q, the countable union of one-point subsets, is of the first category.

7. Q is clearly separable and $R - Q$ is separable since the irrationals of the form $\pi + q$, $q \in Q$ are dense in $R - Q$. Thus Q and $R - Q$ are second countable.

8. Since in either Q or $R - Q$ the only compact sets are nowhere dense, it follows that neither Q nor $R - Q$ are locally compact or σ-locally compact. However Q, being countable, is σ-compact.

9. Both Q and $R - Q$ are totally separated, but since Q and $R - Q$ are both dense-in-themselves neither is scattered, though both are zero dimensional.

10. $[0,1] \cap Q$ is totally bounded, but not compact.

32. Special Subsets of the Real Line

If R is the real line with the Euclidean topology, we consider the following subsets:

1. Let A be the set of all points $1/n$, for $n = 1, 2, 3, \ldots$
   (a) $\bar{A} = A \cup \{0\}$.
   (b) 0 is a limit point and an ω-accumulation point but not a condensation point of the uncountable set $A \cup [2,3]$.
   (c) The set A contains a Cauchy sequence $(1, \frac{1}{2}, \frac{1}{3}, \ldots)$ which has no limit point in A.

2. Let $A = \{0\} \cup \{1/n | n = 1, 2, 3, \ldots\}$.
   (a) A is not locally connected, for no neighborhood of 0 is connected.
   (b) If B is any countable discrete space, and $f \colon B \to A$ any one-to-one correspondence, then B is locally connected and f is continuous, but $A = f(B)$ is not locally connected.
   (c) A is totally separated, since if $a,b \in A$ where $a < b$, we may select an irrational α such that $a < \alpha < b$ and $A \cap [0,\alpha)$ and $A \cap (\alpha,1]$ separates A so both the components and quasicomponents of A are single points.
   (d) A is not extremally disconnected since $\{1/2k\}$ is open, but its closure contains 0 and is not open.

3. Let $\{a_n\}$ be the sequence $1/1$, $1 + 1/1$, $1/2$, $1 + 1/2$, $1/3$, $1 + 1/3$, $\ldots$, $1/m$, $1 + 1/m$, $\ldots$. 0 is a limit point and an ω-accumulation point of the set of numbers in the sequence. It is an accumulation point but not a limit point of $\{a_n\}$.

4. Let $A_n = (1 - 1/n, 1 + 1/n)$ for $n = 1, 2, 3, \ldots$. $\cap A_n =$

$\bigcap A_n{}^\circ = 1$. $(\bigcap A_n)^\circ = \{1\}^\circ = \emptyset$, so $(\bigcap A_n)^\circ$ is properly contained in $\bigcap A_n{}^\circ$.

5. Let $A = (0,\frac{1}{2}) \cup (\frac{1}{2},1)$.
   (a) $A = A^\circ$, $A^{\circ-\circ} = A^{-\circ} = (0,1)$, and $A^{\circ-} = A^{-\circ-} = A^- = [0,1]$.
   (b) $A^{ee} = (0,1)$, so A° is a proper subset of A^{ee}.
   (c) $(0,\frac{1}{2})$ and $(\frac{1}{2},1)$ are both regular open, but their union, A, is not regular open.
   (d) If $B_1 = (0,\frac{1}{2})$ and $B_2 = (\frac{1}{2},1)$, $\overline{(B_1 \cap B_2)} = \emptyset$, but $\bar{B}_1 \cap \bar{B}_2 = \{\frac{1}{2}\}$.

6. If $A_1 = [0,\frac{1}{2}]$ and $A_2 = [\frac{1}{2},1]$, $\{\frac{1}{2}\}$ is the intersection of the regular closed sets A_1 and A_2, but is not itself regular closed. A_1 and A_2 are also an example of sets for which $A_1{}^\circ \cup A_2{}^\circ \neq (A_1 \cup A_2)^\circ$, since $A_1{}^\circ \cup A_2{}^\circ = (0,\frac{1}{2}) \cup (\frac{1}{2},1)$, but $(A_1 \cup A_2)^\circ = (0,1)$.

7. Let $A = (0,1)$ and $B = [0,1]$. A is homeomorphic to the subset $(0,1)$ of $[0,1]$, B is homeomorphic to the subset $[1/4,3/4]$ of A, but A and B are not homeomorphic.

8. Let $A = \{0\} \cup [1,2] \cup \{3\}$ and let $B = [0,1] \cup \{2\} \cup \{3\}$. A and B are homeomorphic as subspaces, but there is no homeomorphism of R onto R taking A onto B.

9. Consider the set $A = \{1/n | n = 1, 2, 3, \ldots\} \cup (2,3) \cup (3,4) \cup \{4\frac{1}{2}\} \cup [5,6] \cup \{x | x \text{ is rational and } 7 \leq x < 8\}$. There are 14 distinct sets that can be formed from A (including A itself) by successive applications of the closure and complement operations; these sets are depicted graphically in Figure 12.

10. Let Z^+ be the set of positive integers. If $d(x,y) = |x - y|$ is the usual metric for Z^+, we define a new metric on Z^+ by $\delta(x,y) = |x - y|/xy$. The metric topologies for (Z^+,d) and (Z^+,δ) are both discrete and thus are homeomorphic. Clearly, every Cauchy sequence in (Z^+,d) must eventually be constant and so (Z^+,d) is complete. The sequence $1, 2, 3, \ldots,$ is a Cauchy sequence in (Z^+,δ) since for $\epsilon > 0$ if we choose an integer $N(\epsilon) > 1/\epsilon$ then for $m,n > N(\epsilon)$ we have $\delta(m,n) < \epsilon$. But clearly $1, 2, \ldots$ cannot converge in (Z^+,δ) and thus (Z^+,δ) is not complete.

33. Special Subsets of the Plane

1. Let $R^2 = R \times R$ be the Euclidean plane. The set A of all points $\{(x,y) | xy \geq 1\}$, where $x,y \in R$, is a closed subset of R^2. The projection map ρ: $R^2 \to R$ taking (x,y) in R^2 to x in R, is open, but it is not closed since $\rho(A)$ is not closed.

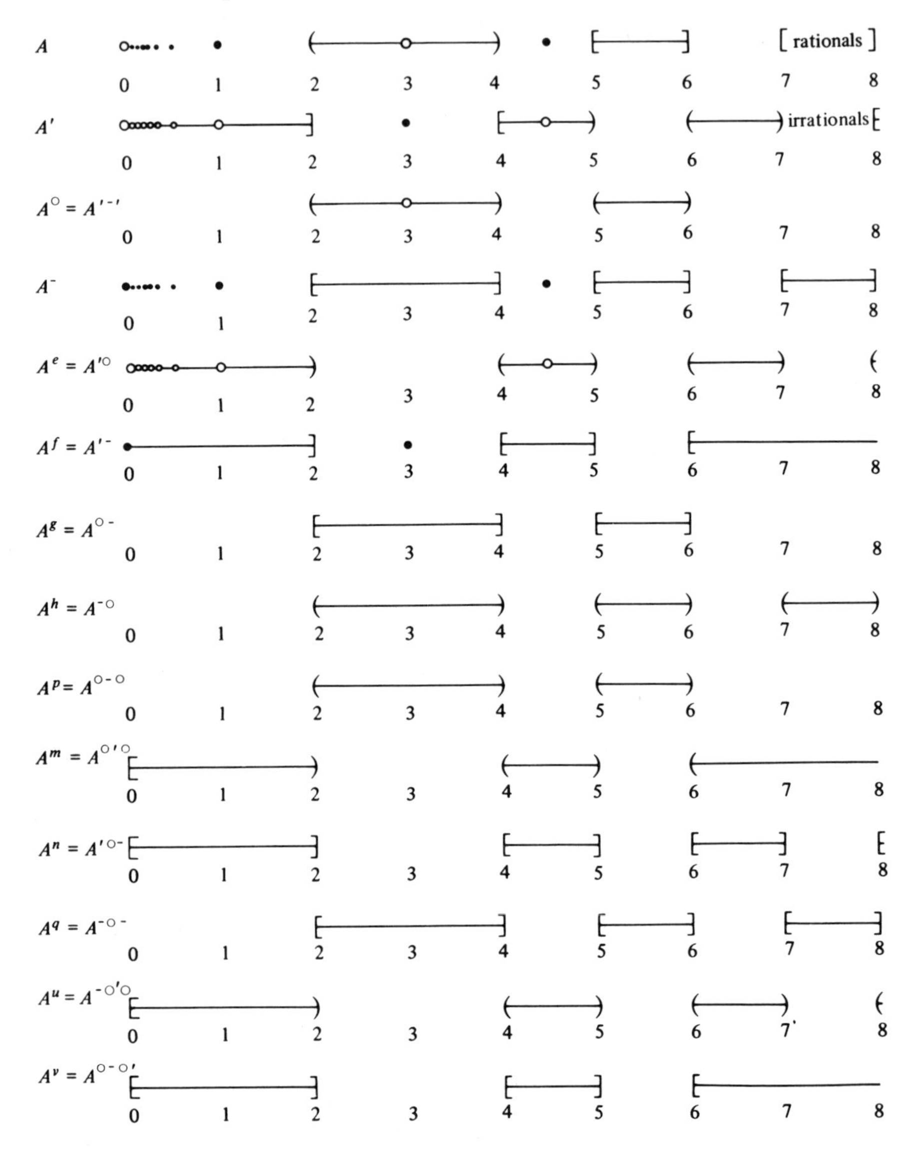

Figure 12.

2. Let A be the subset of $R^2 = R \times R$ consisting of all points with at least one irrational coordinate, and let A have the induced topology. A is arc connected since a point (x_1,y_1) with two irrational coordinates may be joined by an arc to any point $(a,b) \in A$ as follows. Since $(a,b) \in A$ either a or b is irrational, say a. Then

the union of the lines $x = a$, $y = y_1$ is an arc connected subset of X containing (x_1,y_1) and (a,b). Hence any point of A can be connected by an arc to (x_1,y_1).

34. One Point Compactification Topology

35. One Point Compactification of the Rationals

Let (X,τ) be a nonempty topological space, and let p be a point not in X. We define $X^* = X \cup \{p\}$ and describe a topology τ^* on X^* by calling a set in X^* open iff it is in τ or is the complement of a closed and compact subset of (X,τ).

1. (X^*,τ^*) is compact since any open cover $\mathcal{U}$ of X^* contains an open set about p and the complement of this open set is covered by a finite number of sets in $\mathcal{U}$.

2. (X^*,τ^*) is T_1 iff (X,τ) is since if $x \in X^*$, $x \neq p$ then $X^* - \{x\}$ is an open set containing p when (X,τ) is T_1. Conversely (X,τ) is a subspace of (X^*,τ^*) hence is T_1 when (X^*,τ^*) is.

3. (X^*,τ^*) is T_2 iff (X,τ) is T_2 and locally compact. For if $x,y \in X$ they may clearly be separated in X^*; if $y = p$ and (X,τ) is T_2 then any compact neighborhood of x is closed so its complement is an open set about p disjoint from some open set about x. Conversely if (X^*,τ^*) is T_2 so is (X,τ) as a subspace. To see that (X,τ) is locally compact let $x \in X$, and let U be an open neighborhood of x disjoint from an open neighborhood V of p. Then $X^* - V$ is a compact neighborhood of x. Thus since (X^*,τ^*) is compact (X^*,τ^*) is T_4 if it is T_2.

4. If (Q,τ) is the rationals with the topology induced by the Euclidean topology on the reals, (Q^*,τ^*) is not Hausdorff, since (Q,τ) is not locally compact. But since (Q,τ) is T_1 so is (Q^*,τ^*) and thus (Q^*,τ^*) satisfies no higher separation axioms.

5. p is a dispersion point of Q^* for, clearly, $Q^* - \{p\}$ is totally disconnected and Q^* is connected since any open set containing p has a nowhere dense, and thus non open complement. Thus (Q^*,τ^*) is biconnected.

6. Every sequence in (Q^*,τ^*) must either be contained in a compact subset, or must contain a subsequence converging to p. In either case, the original sequence must contain a convergent subsequence, so (Q^*,τ^*) is sequentially compact.

36. Hilbert Space

37. Fréchet Space

Hilbert space H is the set of all sequences $x = \langle x_i \rangle$ of real numbers x_i such that Σx_i^2 converges together with a topology generated by the metric $d(x,y) = [\Sigma(x_i - y_i)^2]^{1/2}$.

1. H is a complete metric space, for whenever $x^1, x^2, x^3, \ldots$ is a Cauchy sequence in H, then for each i, $\{x_i^j\}_{j=1}^{\infty}$ is a Cauchy sequence in the complete metric space R and thus converges to a point of R, say x_i. Then if $x = \langle x_i \rangle$, the points $x - x^j$ eventually belong to H, so $x = (x - x^j) + x^j$ must be in H, and $d(x,x^j) \to 0$.

2. H is separable since the set of all points having finitely many rational coordinates and the rest zero is a countable dense subset. Thus, since H is a metric space it is also second countable and Lindelöf.

3. H is not locally compact since the closed balls $\bar{B}(x,\epsilon) = \{y | d(x,y) \leq \epsilon\}$ are not compact. For the points $y_n = \langle x_1, x_2, \ldots, x_{n-1}, x_n + \epsilon, x_{n+1}, \ldots \rangle$ are in $\bar{B}(x,\epsilon)$, yet $d(y_n,y_m) = \sqrt{2}\epsilon$ whenever $n \neq m$. Thus $\{y_n\}$ has no convergent subsequence.

4. Since H is Hausdorff any compact subset C is closed. If a set C has nonempty interior it is not compact since H is not locally compact at any point. Thus any compact subset of H is nowhere dense. Hence, since H is a complete metric space and thus second category, H is not σ-compact.

5. H is arc connected since the entire line segment joining any two points of H lies entirely in H. That is, if $x = \langle x_i \rangle$ and $y = \langle y_i \rangle$ are in H, then the function f: $[0,1] \to H$ defined by $f(t) = tx + (1 - t)y = \langle tx_i + (1 - t)y_i \rangle$ is a path joining x to y, since $\Sigma(tx_i + (1 - t)y_i)^2$ converges.

6. H is homeomorphic to R^ω, the countable infinite product of copies of the real numbers R.

7. A direct comparison of the corresponding basis elements shows that the product topology on R^ω may be given also by the Fréchet product metric:

$$d(x,y) = \sum \frac{2^{-i}|x_i - y_i|}{1 + |x_i - y_i|}.$$

In this case we call the metric space (R^ω,d) Fréchet space, and can prove, as with Hilbert space, that Fréchet space is complete.

8. Every separable metric space (X,d) can be mapped homeomorphically onto a subspace of Fréchet space by the function $f(x) = \langle d(x,x_i)\rangle$ where $x \in X$ and $\{x_i\}$ is a countable dense subset of X.

38. Hilbert Cube

The subspace I^ω of Hilbert space consisting of all points $x = \langle x_1, x_2, x_3, \ldots\rangle$ such that $0 \leq x_j \leq 1/j$ (or homeomorphically, $|x_j| \leq 1/j$ for each integer j) is known as the Hilbert cube.

1. I^ω is homeomorphic to the countable infinite product of the closed unit interval, $I = [0,1]$. $f : I^\omega \to \prod_{i=1}^{\infty} I_i$ by $f\langle x_1, x_2, x_3, \ldots\rangle = (x_1, 2x_2, 3x_3, \ldots)$ is a bijective function which is both open and continuous. (This is why the Hilbert cube is denoted by I^ω.)
2. I^ω, being a subspace of Hilbert space, is a metric space and thus completely normal.
3. I^ω is separable and second countable, for the points with rational coordinates for a finite number of x_i and 0 for the other x_i form a countable dense subset.
4. I^ω is compact since it is homeomorphic to $\prod_{i=1}^{\infty} I_i$, which is compact by the Tychonoff theorem. This may also be proved directly by considering a sequence $\{x_i\}$ of points of I^ω. The sequence of first coordinates, $\{x_1{}^i\}$ consists of real numbers from the compact interval $[0,1]$, so there is a subsequence of $\{x^i\}$ whose first coordinates converge to some point $x_1 \in [0,1]$. Similarly, the second coordinates of this subsequence belong to $[0,\frac{1}{2}]$, so there must exist another subsequence whose second coordinates converge to a number $x_2 \in [0,\frac{1}{2}]$. We may use induction to continue this process of constructing subsequences. Then the diagonal subsequence consisting of the first member of the first subsequence, the second member of the second subsequence, and so on, converges to the sequence $\langle x_1, x_2, x_3, \ldots\rangle$, which belongs to I^ω.
5. I^ω is arc connected for if $x = \langle x_i\rangle$ and $y = \langle y_i\rangle$ belong to I^ω then so do the elements $tx + (1 - t)y = \langle tx_i + (1 - t)y_i\rangle$ for all $0 \leq t \leq 1$. Similarly, each metric ball contains the entire line segment joining its center to any point in it (since for $0 \leq t \leq 1$, $d(x,tx + (1 - t)y) \leq d(x,y)$). So I^ω is locally arc connected.

39. Order Topology

Let X be a set which is linearly ordered by the transitive relation "$<$." We define the order (or interval) topology on X by taking as a subbasis the rays $\{x|y < x\}$ and $\{x|x < z\}$. Then the intervals $(y,z) = \{x \in X|y < x < z\}$ for each pair $y,z \in X$ where $y < z$ are open.

1. We will call a set $S \subset X$ convex whenever it contains all points which lie between any two of its points: if $a,b \in S$ and if $a < t < b$, then $t \in S$. This concept is to be distinguished from an interval in X which is a set of points lying between two fixed points of x; as usual we denote intervals by (a,b), $[a,b)$, $(a,b]$, or $[a,b]$ according to whether they do or do not contain their endpoints. Clearly every interval is convex but not conversely.

2. The union of any collection of convex sets with nonempty intersection is convex. So any subset S of X can be uniquely expressed as a union of disjoint, nonempty, maximal convex sets called convex components; the component of S which contains the point $p \in S$ is just the union of all convex subsets of S which contain p.

3. Suppose A and B are separated subsets of X; let $A^* = \bigcup\{[a,b]|a,b \in A, [a,b] \cap \bar{B} = \emptyset\}$, and let $B^* = \bigcup\{[a,b]|a,b \in B, [a,b] \cap \bar{A} = \emptyset\}$. Then $A \subset A^*$ since for $a \in A$, $[a,a] = \{a\}$ is disjoint from $\bar{B}$. Further $A^* \cap B^* = \emptyset$, for if $p \in A^* \cap B^*$, then there must be points $a,b \in A$ and $c,d \in B$ such that $p \in [a,b] \cap [c,d]$. But since neither c nor d can belong to $[a,b]$, and neither a nor b to $[c,d]$, we must have $[a,b] \cap [c,d] = \emptyset$.

4. In fact, we can prove more: A^* and B^* must be separated. To prove this, we observe first that $\overline{A^*} \subset A^* \cup \bar{A}$. For suppose $p \notin A^* \cup \bar{A}$. Then there exists an open interval (s,t) disjoint from A but containing p. The interval (s,t) may intersect A^* only if it intersects some interval $[a,b] \subset A$ where $a,b \in A$. But since $(s,t) \cap A = \emptyset$ and $a,b \in A$ then $(s,t) \subset (a,b)$ which would imply that $p \in A^*$. But since $p \notin A^*$, we must have $(s,t) \cap A^* = \emptyset$. Thus $p \notin \overline{A^*}$. Thus $\overline{A^*} \cap B^* \subset (A^* \cup \bar{A}) \cap B^* = (A^* \cap B^*) \cup (\bar{A} \cap B^*) = \emptyset$.

5. If we now write A^*, B^* and $(A^* \cup B^*)'$ as the union of convex components, $A^* = \bigcup A_\alpha$, $B^* = \bigcup B_\beta$ and $(A^* \cup B^*)' = \bigcup C_\gamma$, the collection $M = \{A_\alpha, B_\beta, C_\gamma\}$ inherits a linear order from X and is thus itself a linearly ordered set. We claim that in the ordered set M, each of the sets A_α (and similarly, each of the sets B_β) has an immediate successor whenever A_α intersects the closure of S_α,

the set of strict upper bounds for A_α. In this case we can show that the successor to A_α is an element of $\{C_\gamma\}$, which we will denote by $C_\alpha{}^+$.

For suppose $A_\alpha \cap \bar{S}_\alpha \neq \emptyset$; then $A_\alpha \cap \bar{S}_\alpha$ contains precisely one point, say p, which belongs to the complement of the closed set $\overline{B^*}$, so there exists a neighborhood (x,y) of p disjoint from $\overline{B^*}$. Then $(x,y) \cap S_\alpha \neq \emptyset$, so $(p,y) \neq \emptyset$. But (p,y) is disjoint from both A^* and B^*, so there must exist some set C_γ containing (p,y). In the linear order on M, C_γ is the immediate successor to A_α, and we will call it $C_\alpha{}^+$.

6. For each γ, select and fix some point $k_\gamma \in C_\gamma$. Then whenever $A_\alpha \cap \bar{S}_\alpha \neq \emptyset$, there exists a unique $k_\alpha{}^+ \in C_\alpha{}^+$, the immediate successor of A_α. In such cases, let $I_\alpha = [p,k_\alpha{}^+)$ where $p \in A \cap \bar{S}_\alpha$; otherwise, if $A \cap \bar{S}_\alpha = \emptyset$, let $I_\alpha = \emptyset$. Define J_α similarly for the strict lower bounds of A_α (using the same collection of points $k_\gamma \in C_\gamma$). Then for each α, let $U_\alpha = J_\alpha \cup A_\alpha \cup I_\alpha$, and similarly for each β, let $V_\beta = J_\beta \cup B_\beta \cup I_\beta$. Each U_α and V_β is clearly a convex open set containing A_α and B_α, respectively. Thus $U = \bigcup U_\alpha$, $V = \bigcup V_\beta$ are open sets containing A^* and B^*, respectively. Since no A_α intersects any B_β, and the use of the same k_γ throughout implies that no J_β or I_β may intersect any J_α or I_α, it is clear that no U_α can intersect any V_β. Thus $U \cap V = \emptyset$, and hence X is T_5. Since the points of X are clearly closed, X is T_1, and thus completely normal.

7. The order topology on X is compact iff it is complete—that is, iff every nonempty subset of X has a greatest lower bound and a least upper bound. This condition is clearly necessary, for if $A \subset X$ and if A has no least upper bound, then the sets $P_\alpha = \{x|x < \alpha\}$ and $S_\beta = \{x|x > \beta\}$ for $\alpha \in A$ and β an upper bound of A cover X but they contain no finite subcover. To prove it sufficient we need only consider, for any given open cover $\mathfrak{U}$ of X, the set S of those elements $y \in X$ for which $[a,y)$ (where $a =$ g.l.b. X) can be covered by finitely many members of $\mathfrak{U}$. If $\alpha =$ l.u.b. S and if $\alpha \in U \in \mathfrak{U}$, then $U \subset S$. There then exists, unless $\alpha =$ l.u.b. X, an interval $(x,y) \subset U$ such that $\alpha \in (x,y)$. Then $(\alpha,y) = \emptyset$ since $\alpha =$ l.u.b. S. But this would mean that $y \in S$, which is impossible. Thus $S = X$.

8. Whenever X contains two consecutive points (that is, whenever some interval (a,b) is empty), X can be separated by $\{x|x \leq a\}$ and $\{x|x \geq b\}$. Similarly, if X contains a bounded set A with no

least upper bound, the set of upper bounds for A and its complement separate X. Thus X is connected only if it contains no consecutive points, and if every bounded subset has a least upper bound. These conditions are in fact sufficient; and are often summarized by the Dedekind cut axiom: if A and B are disjoint nonempty subsets of X whose union is X, and if every point of A is less than every point of B, then there exists l.u.b. A and it equals g.l.b. B. We will use this version to prove sufficiency.

Suppose U and V are disjoint nonempty open sets whose union is X, and assume that U contains a point u which is less than some point $v \in V$. Let E be the convex component of U which contains u, and let $A = E \cup \{x \in X | x < u\}$. If $B = X - A$, then $v \in B$ and so the Dedekind cut axiom guarantees the existence of a point $p = \text{l.u.b. } A = \text{g.l.b. } B$. If $p \in A$, then it must be in E and thus in U; so there exist points x,y such that $p \in (x,y)$, $(x,y) \subset E \subset U$. But since $p = \text{l.u.b. } A$, $(p,y) = \emptyset$ which is impossible since y cannot be the immediate successor of p. Thus $p \notin A$. By a similar argument it can be seen that $p \notin B$, which gives the desired contradiction.

9. If X is a connected set with the order topology, any point $p \in X$ is a cut point, since $X - \{p\}$ is separated by $P_p = \{x \in X | x < p\}$ and $S_p = \{x \in X | x > p\}$.

40. Open Ordinal Space $[0,\Gamma)$ $(\Gamma < \Omega)$

41. Closed Ordinal Space $[0,\Gamma]$ $(\Gamma < \Omega)$

42. Open Ordinal Space $[0,\Omega)$

43. Closed Ordinal Space $[0,\Omega]$

Closed ordinal space consists of the set of all ordinal numbers less than or equal to some limit ordinal Γ, together with the order topology. Open ordinal space is the subspace $[0,\Gamma)$ consisting of all ordinals strictly less than Γ. Sets of the form $(\alpha,\beta + 1) = (\alpha,\beta] = \{x | \alpha < x < \beta + 1\}$ form a basis for this topology. We will consider two special cases: $\Gamma = \Omega$, the first uncountable ordinal, and $\Gamma < \Omega$.

1. In ordinal space $[0,\Omega]$, $\{\Omega\}$ is a closed set that is not a G_δ-set. $\{\Omega\}$ is closed since its complement $[0,\Omega)$ is an open set. It is not a G_δ-set, since for any countable collection G_i of open sets containing Ω, we can find a collection of basis elements of the form

$(\alpha_i,\Omega] \subset G_i$ for each i. The least upper bound of the α_i is an ordinal γ less than Ω, since each α_i, or equivalently, each $[0,\alpha_i)$ is countable and the countable union of countable sets is countable. Therefore $\cap G_i \supset (\gamma,\Omega] \neq \{\Omega\}$.

2. Thus ordinal space $[0,\Omega]$ is not first countable, since the point Ω does not have a countable local basis. In fact, Ω is a limit point of the set (α,Ω) but it is not the limit point of any sequence of points in (α,Ω).

3. Similar reasoning shows that $[0,\Omega)$ is not separable, for the least upper bound of any countable subset of $[0,\Omega)$ is countable, and will be strictly less than Ω. Therefore, there will always be an open interval (α,Ω) in the complement of a countable subset. Thus both $[0,\Omega)$ and $[0,\Omega]$ fail to be separable. But unlike $[0,\Omega]$, $[0,\Omega)$ is first countable, since the only point of $[0,\Omega]$ which does not have a countable local basis is Ω.

4. Since all order topologies satisfy all the separation axioms, each ordinal space is completely normal. But $[0,\Omega]$ is not perfectly normal, since the closed set $\{\Omega\}$ is not G_δ.

5. Although neither $[0,\Omega]$ nor $[0,\Omega)$ are second countable, both $[0,\Gamma]$ and $[0,\Gamma)$ are (for $\Gamma < \Omega$) since each point has a countable local basis, and there are only countably many points. Thus, since ordinal spaces are regular, both $[0,\Gamma]$ and $[0,\Gamma)$ are metrizable.

6. Every subset of each ordinal space has a greatest lower bound (its first element) and every subset of $[0,\Gamma]$ has a least upper bound. Therefore, $[0,\Gamma]$ is a complete order topology, and thus compact. Similarly, the closure of each basis neighborhood is compact, so every ordinal space is strongly locally compact.

7. The open subset $[0,\Gamma)$ of $[0,\Gamma]$ fails to be compact since the collection $\{[0,\alpha)|\alpha < \Gamma\}$ is an open cover with no finite subcover (since Γ is a limit ordinal).

8. Since $[0,\Omega]$ is compact, it is countably compact. Thus every sequence in $[0,\Omega)$ has an accumulation point in $[0,\Omega]$. But Ω cannot be an accumulation point of any sequence in $[0,\Omega)$. So every sequence in $[0,\Omega)$ has an accumulation point in $[0,\Omega)$, which means that $[0,\Omega)$ is countably compact.

9. Because a space is compact iff it is both countably compact and metacompact, and since $[0,\Omega)$ is countably compact but not compact, $[0,\Omega)$ cannot be metacompact or paracompact.

10. Since every T_3 Lindelöf space is paracompact, $[0,\Omega)$ is not Lindelöf, and thus not σ-compact. But $[0,\Omega]$ being compact, is both Lindelöf and σ-compact.

11. Since it is first countable and countably compact, $[0,\Omega)$ is sequentially compact.

12. Any continuous real-valued function f on $[0,\Omega)$ must be eventually constant—that is, constant on some set (α,Ω). To prove this, we verify first the existence of a sequence $\alpha_n \in [0,\Omega)$ such that $|f(\beta) - f(\alpha_n)| < 1/n$ whenever $\beta > \alpha_n$. For if no such sequence existed, there would be some integer n_0 for which we could construct inductively an increasing sequence $\gamma_i \in [0,\Omega)$ such that $|f(\gamma_i) - f(\gamma_{i-1})| \geq 1/n_0$ for each i. But the sequence γ_i converges to its least upper bound γ, whereas the points $f(\gamma_i)$ cannot converge. This is impossible for a continuous function f. So the sequence α_n exists, and it has a least upper bound $\alpha < \Omega$. Clearly f is constant on (α,Ω).

13. All ordinal spaces are zero dimensional, since the basis elements $(\alpha,\beta]$ are closed. But none of them are extremally disconnected, since the open set $A = \{1, 3, 5, 7, \ldots\}$ has as its closure $A \cup \{\omega\}$, a set which is not open.

14. Since ordinals are well-ordered, every subset of ordinal space with at least two points contains both a first and a second element; thus the first element cannot be a limit point of the set, so no nonempty subset of ordinal space can be dense-in-itself. Thus each ordinal space is scattered.

15. The ordinal space $[0,2\Omega)$ is, like $[0,\Omega)$, sequentially compact but not compact. But, unlike $[0,\Omega)$, $[0,2\Omega)$ is not first countable.

16. If we expand the interval topology on $[0,\omega\Omega)$ by declaring open each ordinal $n\Omega$, we will have, essentially, a countably infinite sum of copies of $[0,\Omega)$. This new space will fail, like $[0,\Omega)$, to be metacompact and furthermore it will fail to be countably compact since the summands form a countable cover with no finite subcover. But clearly $[0,\omega\Omega)$ with the new topology will still be countably paracompact.

44. Uncountable Discrete Ordinal Space

Let X be the set of points of the form $\alpha + 1$ in $[0,\Omega)$ where α is a limit ordinal, together with the subspace topology induced by the order topology on $[0,\Omega)$.

1. X is well-ordered, since it is a subset of $[0,\Omega)$ yet its topology is discrete since $X \cap (\alpha,\alpha + 2) = \{\alpha + 1\}$ is open in X. Thus the topology on X is not the topology given by its natural ordering.
2. The product set $[0,\Omega) \times Z$, where Z is the integers with the lexicographic order $((\alpha,n) < (\alpha',n')$ if $\alpha < \alpha'$ or if $\alpha = \alpha'$ and $n < n')$ is a linearly ordered set in which every element has both an immediate predecessor and an immediate successor. Thus its order topology is also discrete and so, since it has the same cardinality as X, it is homeomorphic to X.
3. The discrete metric on X yields the discrete uniformity U_1 which has as a base the diagonal $\Delta = \{(x,x) \in X \times X | x \in X\}$. Thus U_1 is a metrizable uniformity.
4. The uniformity U_2 which is generated by the basis of all sets of the form $B_z = \Delta \cup \{(x,y) \in X \times X | x > z$ and $y > z\}$ also yields the discrete topology. However U_2 does not have a countable base since every countable subset of X has a least upper bound less than Ω. Thus (X,U_2) is a nonmetrizable uniform space whose topology is metrizable.

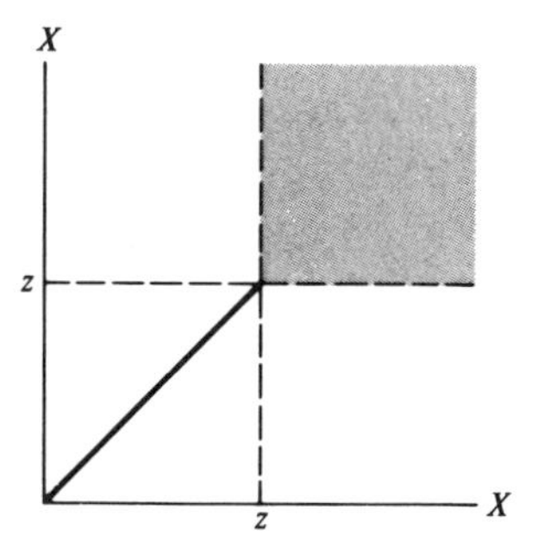

45. The Long Line

46. The Extended Long Line

The long line L is constructed from the ordinal space $[0,\Omega)$ (where Ω is the least uncountable ordinal) by placing between each ordinal α and its successor $\alpha + 1$ a copy of the unit interval $I = (0,1)$. L is then linearly ordered, and we give it the order topology. The extended long line L^* is constructed similarly from $[0,\Omega]$.

1. L is not compact, since the open covering by sets of the form $\{y | y < \alpha\}$, $\alpha \in [0,\Omega)$ has no finite subcover. In fact, it has no countable subcover, since the least upper bound of any countable collection of ordinals $\alpha \in [0,\Omega)$ must be countable, and therefore cannot equal Ω. Thus L is not Lindelöf, and therefore not σ-compact.
2. For a similar reason, L is not separable. If D is a countable subset of L, and if β is the least upper bound of D, the set $\{y \in L | y > \beta\}$ is a nonempty open subset of L which is disjoint from D. So D cannot be dense in L.

3. L is first countable since each point $x \in L$ is the least upper bound of a countable collection $\{x_n\}$ of points in L which precede it. So $\{(x_n,x + 1/n)\}$ is a countable local basis for the topology at the point x.

4. Since L and L^* carry the order topology, they are both completely normal. However neither L nor L^* is perfectly normal since the sets of limit ordinals in either space are closed but not G_δ. For example, let $\{U_i\}$ be a countable collection of open sets containing the set S of limit ordinals in L. Let f_i be a Urysohn function for S and the complement of U_i. Then f_i must be constant for all values greater than some α_i. Hence $\bigcap U_i$ contains all points of L greater than the least upper bound of the α_i. So $\bigcap U_i \neq S$.

5. L^* is compact since each open neighborhood of Ω has a compact complement. So L is countably compact, for just as in the ordinal space $[0,\Omega)$, every sequence in L has an accumulation point which must be in L. Clearly L is not compact, and thus, since it is countably compact, neither metacompact nor paracompact.

6. L is arc connected since whenever $p,q \in L$, the interval $[p,q]$ is homeomorphic to the closed unit interval. L^*, being the closure of L, is thus connected, but it is not path connected, for no path can join any point to Ω.

7. Since both L and L^* are countably compact and regular, they are of the second category.

47. An Altered Long Line

To the long line L we add a point p. Open sets of $L \cup \{p\}$ are the open sets of L together with those generated by the following neighborhoods of p: $U_\beta(p) = \{p\} \cup \{\bigcup_{\alpha=\beta}^{\Omega} (\alpha,\alpha + 1)\}$ (where $1 \leq \beta < \Omega$). $U_\beta(p)$ is then a right-hand ray less the ordinal points. We consider p to be the greatest element of $L \cup \{p\}$.

1. If $a,b \in L \cup \{p\}$, say $a < b$, then there exists a z such that $a < z < b$. Thus $\{x|x < z\}$ and $\{x|x > z\}$ are disjoint neighborhoods of a and b, so $L \cup \{p\}$ is T_2.

2. No $U_\beta(p)$ contains $\overline{U_\alpha(p)}$ for any α. Hence $L \cup \{p\}$ is not T_3;

but since it is T_2, it cannot be compact. Yet, $L \cup \{p\}$, like the extended long line, is countably compact, connected, and not separable. But it is not Lindelöf, for the cover consisting of the sets $(\beta,\beta + 2)$ together with $U_1(p)$ has no countable subcover.

48. Lexicographic Ordering on the Unit Square

Let X be the unit square in the plane: $X = \{(x,y)|0 \leq x \leq 1, 0 \leq y \leq 1\}$. We order X lexicographically ($(x,y) < (u,v)$ iff $x < u$, or $x = u$ and $y < v$) and place the order topology on X.

1. Since X carries the order topology, it is completely normal.

2. Every nonempty subset A of X has a least upper bound. Although this is not obvious, it is at least clear that the set of first coordinates of points of A, being a subset of the closed unit interval, has a least upper bound; let $\alpha = \text{l.u.b. } \{(x,0)|(x,y) \in A \text{ for some } y\}$. Then if $A \cap \{(\alpha,y)|0 \leq y \leq 1\} = \emptyset$, $(\alpha,0)$ is the least upper bound for A. Otherwise, $\text{l.u.b. } A = \text{l.u.b. } \{(\alpha,y) \in A|0 \leq y \leq 1\}$. Thus X is a complete ordered space, and hence compact.

3. The set $L = \{(x,y)|y = \frac{1}{2}\}$ is an uncountable discrete subspace because each of its points can be obtained by intersecting L with the open sets $U_x = \{(x,y)|\frac{1}{4} < y < \frac{3}{4}\}$. The family $\{U_x|0 \leq x \leq 1\}$ is an uncountable collection of disjoint open sets, so X does not satisfy the countable chain condition. Hence X is not separable.

4. Since X is compact but not separable, it is not metrizable. However, it is first countable.

5. Since in the linear order on X there are no consecutive points, and since every (bounded) subset of X has a least upper bound, X is connected. But X is not path connected since any path in X joining, say, (0,0) and (1,1) must be connected, and therefore must contain all of X—since in a linearly ordered space any connected set containing two given points must contain the entire interval between them. But X cannot be the continuous image of [0,1] since X contains an uncountable collection of disjoint open sets whose inverse images would form an uncountable collection of disjoint open sets in [0,1]. But this is impossible since each such open set would have to contain a rational.

49. Right Order Topology

50. Right Order Topology on *R*

If X is a linearly ordered set, the topology generated by basis sets of the form $S_a = \{x|x > a\}$ is called the right order topology on X. (A left order topology is defined similarly using the sets $P_a = \{x|x < a\}$.)

1. For any point $a \in X$, every $x < a$ is a limit point for $\{a\}$. So the closure of any open set is the whole space X, and every right order topology is weakly countably compact.

2. Clearly X is both hyperconnected and ultraconnected, thus path connected, locally connected, and pseudocompact.

3. X is always locally compact, but it is compact iff it contains a first element. But since the closure of any open set is the whole space, X will be strongly locally compact iff X is compact.

4. If $x < y$, then S_x is an open neighborhood of y which does not contain x. So X is T_0, but clearly not T_1. Thus it is not T_2, T_3, or $T_{3\frac{1}{2}}$. But since there are no disjoint closed sets, X is T_4 vacuously. Similarly, X is T_5 vacuously since there can be no separated sets in X: if A and B are disjoint and nonempty, one of them, say A, must contain a point p which is greater than some point $q \in B$. But then $q \in \bar{A}$, since every neighborhood of q must contain p. Thus $\bar{A} \cap B \neq \emptyset$, so A and B cannot be separated.

5. X is not perfectly T_4, since the only open set which contains any closed set is X.

6. An interesting special case is the right order topology on the real numbers R; call this space (R,τ). Then (R,τ) is second countable since $\{S_r\}_{r\in Q}$ is a countable basis for τ. Thus (R,τ) is Lindelöf, and therefore not countably compact since it is not compact. But, since (R,τ) is thus both locally compact and Lindelöf, it must be σ-compact.

7. Each set $P_r = \{x \in R|x < r\}$ is nowhere dense in (R,τ), so R, which equals $\bigcup_{r\in Q} P_r$, is first category. But each P_r is dense-in-itself.

8. The open cover $\{S_n\}$ of R, where n is an integer, has no point finite refinement, so (R,τ) is not countably metacompact.

9. Any finite set in (R,τ) has infinitely many limit points, but no ω-accumulation points. Thus if we add to a finite set its ω-accumulation points, we will not produce a closed set.

51. Right Half-Open Interval Topology

On the set X of all real numbers (or, more generally, any linearly ordered set), we choose a basis for a topology τ to be the family of all sets of the form $[a,b)$, where $a,b \in X$. For obvious reasons, (X,τ) is called the right half-open interval topology, or sometimes the lower limit topology. Then sets of the form $(-\infty,a)$, $[a,b)$, or $[a,+\infty)$ are both open and closed. Sets of the form (a,b) or $(a,+\infty)$ are open in X, since $(a,b) = \bigcup\{[\alpha,b) | a < \alpha < b\}$. They are not closed, since sets of the form $(-\infty,a]$, $[a,b]$, and $\{p\}$ are not open, not being the union of basis elements.

1. If $A = (a,+\infty)$, $A^\circ = (a,+\infty)$, $A^e = (-\infty,a)$, and $A^{ee} = [a,+\infty)$. We see that $A^{ee} \supset A^\circ$, but $A^{ee} \neq A^\circ$.

2. X is Hausdorff since the topology is an expansion of the interval topology on X. X is in fact completely normal, for let A and B be two separated sets in X. Then for each $a \in X - \bar{B}$ there is an $x_a \in X$ such that $[a,x_a) \subset X - \bar{B}$ since $X - \bar{B}$ is open. We define $O_A = \bigcup_{a \in A} [a,x_a)$. O_B containing B is defined analogously. If $O_A \cap O_B \neq \emptyset$, then for some $a \in A$, $b \in B$ we have $[a,x_a) \cap [b,x_b) \neq \emptyset$. Say $a < b$: then $b \in [a,x_a) \subset X - \bar{B}$, a contradiction. So O_A and O_B are disjoint.

3. X is not second countable, for if $S = \{[x_i,y_i) | i \in Z^+\}$ is any countable set of basis elements there exists an $a \in X$ such that $a \neq x_i$ for any $i \in Z^+$. Then for any $b > a$, $[a,b)$ is not a union of any collection of elements of S.

4. However X is first countable, for at a point x, the collection of sets of the form $[x,a_i)$, where a_i is a rational number, form a countable local basis. Furthermore, X is separable since the rational numbers are dense in X. Thus, since X is not second countable, it cannot be metrizable.

5. Every compact subset of (X,τ) is countable and nowhere dense in the Euclidean topology. In fact, if A is compact, and if $a \in A$, then there is some interval (x_a,a) disjoint from A. (If not, A would contain an increasing sequence $\{b_i\}$ such that $|b_i - a| \to 0$; then $[-\infty,b_i)$, $\{[b_i,b_{i+1})\}_{i=1}^{\infty}$, and $[a,\infty)$ is a (countable) open covering of A which can have no finite subcover.) Hence, since the real numbers are uncountable, (X,τ) is not σ-compact; neither is it locally compact, for every open set (in τ) is uncountable.

6. But (X,τ) is Lindelöf, for if $\{U_\alpha\}$ is an open covering of (X,τ) and if $U_\alpha{}^\circ$ is the Euclidean interior of U_α, then since every subset of R is Lindelöf, $\{U_\alpha{}^\circ\}$ has a countable subcollection $\{U_{\alpha_i}{}^\circ\}$ which covers $U = \bigcup U_\alpha{}^\circ$. But the complement $A = X - U$ may be covered by a countable subcollection of $\{U_\alpha\}$ since A is a countable set. For if $p \in A$ there must be some point $x_p > p$ such that $(p,x_p) \cap A = \emptyset$. But these intervals are disjoint, so there cannot be uncountably many of them.

7. Since each of the basis sets $[a,b)$ is both open and closed, X is zero dimensional; since it is T_0, it is therefore also totally separated and totally disconnected. But, since X is dense-in-itself, it is not scattered, nor is it extremally disconnected since the closure of the set $\bigcup_{n=1}^{\infty} [1/2n, 1/(2n - 1))$ is not open.

8. Since X is both Lindelöf and regular, it is paracompact and thus fully normal.

52. Nested Interval Topology

On the open interval $X = (0,1)$ we define a topology τ by declaring open all sets of the form $U_n = (0,1 - 1/n)$, for $n = 2, 3, 4, \ldots$, together with $\emptyset$ and X.

1. Since every nonempty open set contains both $\frac{1}{8}$ and $\frac{1}{4}$, X is not T_0, thus not T_1, T_2, or $T_{2\frac{1}{2}}$.

2. Similarly, since every nonempty open set contains $\frac{1}{8}$, every neighborhood of the closed set $[\frac{1}{2},1)$ must also. Thus (X,τ) is not T_3.

3. Since the closure of any set S includes all points of X greater than the greatest lower bound of S, there can be no separated sets; thus (X,τ) is T_5 vacuously, and thus T_4 also vacuously.

4. X is clearly hyperconnected and ultraconnected and thus is path connected, connected, locally connected, and pseudocompact.

5. Since τ is countable, X is second countable and thus first countable, separable, and Lindelöf.

6. Since all nontrivial open sets are of the form $U_n = (0,1 - 1/n)$, each open set except X itself is compact. Furthermore, no closed

set except $\emptyset$ is compact, since $\{U_n\}$ is an open covering of any closed set which can never have a finite subcover.

7. Since each point of X is contained in a compact open set $(0,1 - 1/n)$, X is locally compact. X is not strongly locally compact since the closures of neighborhoods are not compact: $\bar{U}_n = X$ for each n. Since $X = \bigcup U_n$, X is σ-compact and thus σ-locally compact.

8. X is not countably compact for the U_n have no finite subcover, and thus not sequentially compact. X is, however, weakly countably compact for if p is the greatest lower bound of any infinite set A, then every point $x > p$ is a limit point of A. X is not countably metacompact since the open cover $\{U_n\}$ has no finite refinement and 1/4 is in every set in the cover.

53. Overlapping Interval Topology

On the set $[-1,1]$ we generate a topology from sets of the form $[-1,b)$ for $b > 0$ and $(a,1]$ for $a < 0$. Then all sets of the form (a,b) are also open.

1. X is T_0, but not T_1, since the point 0 is not closed. Also X is not T_4 since $\{-1\}$ and $\{1\}$ are closed subsets with no disjoint neighborhoods.

2. X is compact, since in any open covering, the two sets which include 1 and -1 will cover X.

3. Every nonempty open set contains 0, and the closure of any nonempty open set is the whole set X. So X is dense-in-itself and hyperconnected and thus connected and locally connected.

4. Since this topology is coarser than the Euclidean topology X is arc connected.

5. X is second countable since the intervals $[-1,t)$, (s,t) and $(s,1]$ for rational s,t such that $s < 0 < t$ form a countable basis.

6. The sequence $0, \frac{1}{2}, 0, \frac{1}{2}, 0, \frac{1}{2}, \ldots$ has 0 as an accumulation point but not a limit point but any point greater than $\frac{1}{2}$ is a limit point of this sequence.

54. Interlocking Interval Topology

Let $X = R^+ - Z^+$, the positive real numbers excluding the positive integers. The topology τ on X is generated by the sets $S_n = (0,1/n) \cup (n,n + 1)$, where $n \in Z^+$.

1. X is not T_0 since $2\frac{1}{2}$ and $2\frac{1}{4}$ cannot be separated; X is also not T_4 since $(2,3)$ and $(3,4)$ are disjoint closed sets, but X has no disjoint open sets.

2. Since X is hyperconnected it is connected, locally connected, and pseudocompact.

3. $\{S_n\}$ is a countable open covering of X with no finite subcover. So (X,τ) is not countably compact, and thus not compact.

4. A basis for τ consists of the sets S_n together with sets of the form $(0,1/n)$ for $n \geq 2$. Since each of the basis elements is compact, (X,τ) is locally compact and thus σ compact since $\bigcup S_n = X$.

5. Since $\{S_n\}$ is countable, (X,τ) is second countable and therefore separable and Lindelöf.

6. The open cover $\{S_n\}$ has no refinement. But $S_1 = (0,1) \cup (1,2)$ intersects every other set in the cover. So (X,τ) is not countably paracompact.

7. But the cover $\{S_n\}$ is point finite, since each point $x > 1$ belongs to only one member of the cover, and each point $x < 1$ belongs to finitely many members of the cover $\{S_n\}$. So (X,τ) is metacompact.

55. Hjalmar Ekdal Topology

The Hjalmar Ekdal topology is defined on the set X of positive integers by including in τ precisely those subsets of X which contain the successor of every odd integer in them. Thus a set A is closed in (X,τ) iff for each even point p in A, $p - 1 \in A$, for if $p - 1 \notin A$, it would be an odd integer in $X - A$.

1. (X,τ) is just the sum of countably many copies of Sierpinski space.

2. The sum of spaces is T_i iff each of the spaces is T_i since to separate two sets it is necessary and sufficient to separate them within each summand, for the summands are both open and closed. Thus X is T_0, T_4, and T_5 only. In this case the summands are the subsets $\{2n - 1,2n\}$.

3. X is not compact since the covering by summands is an open disjoint infinite covering with no finite subcover. But since each summand is second countable so is the sum X. The fact that the cover by summands is a refinement of any open cover implies that X is paracompact. That these summands are finite implies that X is locally compact.

4. Since the subspaces $\{2n - 1,2n\}$ are the components of X, X is locally connected and locally path connected, but neither totally separated nor totally disconnected. No subset is dense-in-itself, so the space is scattered. In fact, since each even integer is open and not closed, X is neither first category nor zero dimensional.

56. Prime Ideal Topology

Let X be the set of all prime ideals of integers; that is, X is the set of all ideals P in Z whose complement is multiplicatively closed. We define a topology for X by taking as a basis all sets $V_x = \{P \in X | x \notin P\}$, for all $x \in Z^+$. (Note that $V_0 = \varnothing$ and $V_1 = X$.)

1. Each basis element V_x contains all but a finite number of prime ideals (those generated by the prime factors of x), while the ideal 0 is contained in every basis element V_x. Thus a subset of X is open iff it contains 0 and has a finite complement.

2. A subset U of X is open iff there is an ideal $I \subset Z$ such that $U = \{P \in X | I \not\subset P\}$. For if $U = \bigcup_{x \in M} V_x$, I is the ideal generated by M, and conversely. So a set C is closed iff there is an ideal I in Z such that $C = \{P \in X | I \subset P\}$. (This description may be used to define a topology on the set of prime ideals in any ring A; this space is normally called *Spec A*.)

3. X is T_0, but not T_1 since 0 is in every open set. Thus X is not T_2 or T_3. Since any nonzero prime ideal is closed X does have disjoint closed sets and hence is not T_4 or T_5, nor metrizable. Again since 0 is in every open set X is hyperconnected and thus connected and locally connected.

4. X is compact since every open set has a finite complement. Also X is second countable since the basis $\{V_x\}_{x \in Z^+}$ is countable.

5. The map $f\colon [0,1] \to X$ defined by $f(0) = P$, $f(1) = Q$ and $f(t) = 0$ for $t \in (0,1)$ is continuous for any $P,Q \in X$. Thus X is both path connected and locally path connected since 0 is in every neighborhood.

57. Divisor Topology

Let $X = \{x \in Z^+ | x \geq 2\}$, together with the topology generated by sets of the form $U_n = \{x \in Z^+ | x \text{ divides } n\}$, for $n \geq 2$.

1. X is T_0 for if $x < y$ then $y \notin U_x$; but every neighborhood of 6 contains 3 so X is not T_1, and thus not T_2 or T_3.

2. If x is a point of X the closure of x consists of all multiples of x. Thus no two nonempty closed sets are disjoint since they must both contain the product of any two elements, one from each set. Thus X is T_4 vacuously, and also ultraconnected, path connected, and connected.

3. If neither of x or y divides the other they form separated sets, but any open set containing x or y contains the greatest common divisor of x and y. Thus 6 and 8 may not be separated by open sets so X is not T_5.

4. The set of primes is dense in X for every open set contains all prime divisors of all of its elements; in fact each prime is open and therefore not nowhere dense so X is second category.

5. Since each point has a finite neighborhood, X is locally compact. But it is neither countably compact, since the sets $S_n = \bigcup_{x \leq n} U_n = \{x | x \leq n\}$ form an open covering of X with no finite subcover, nor strongly locally compact, since every closed set is infinite and therefore not compact.

6. Since X is countable, it is separable and Lindelöf; and since all basis elements are finite, it is second countable.

7. X is locally connected, since for each $n \in X$ the set U_n is a smallest open set containing n, and thus connected. X, though connected, is scattered since each nonempty subset has a first element which is therefore an isolated point of the set, and therefore no such set can be dense-in-itself.

8. Each basis neighborhood U_n is ultraconnected in the induced topology (since every closed set in U_n contains the point n), so X is locally path connected.

58. Evenly Spaced Integer Topology

Let X be the set of integers with the topology generated by sets of the form $a + kZ = \{a + k\lambda | \lambda \in Z\}$, where Z is the set of integers, and $a,k \in X$, and $k \neq 0$. The basis sets are simply the cosets of nonzero subgroups of the integers.

1. X is Hausdorff, for if $a,b \in X$ and k does not divide $b - a$, then $a + kZ$ and $b + kZ$ are disjoint cosets.

2. Every basis element is closed since a given coset of a subgroup is

the complement of the union of the remaining cosets of that subgroup. Thus X is zero dimensional and regular.

3. X is totally separated, since it is Hausdorff and zero dimensional. If S is the set of odd integers, the sets $2^n S$, $n = 0, 1, 2, \ldots$, are disjoint, for if $a2^n = b2^m$, then $a2^{n-m} = b$ which implies either $m = n$ or one of a and b is even. The sets $A = \bigcup_{n=0}^{\infty} 2^{2n}S$ and $B = \bigcup_{n=0}^{\infty} 2^{(2n+1)}S$ are then open, disjoint, and contain all the integers except 0. If $k \in A$ then $2k \in B$ and vice versa, thus 0 is a limit point of both A and B so X is not extremally disconnected.

4. X is clearly second countable and thus, since it is regular, it is metrizable which in turn implies that X satisfies all the separation axioms and is paracompact. Since points are closed and not open, and since X is countable, X is first category, and thus not complete.

5. The set of positive primes congruent to 3 mod 4 is an infinite set without a limit point, so X is not countably compact and thus not compact. However, since X is countable and second countable, it is σ-compact and Lindelöf.

6. The function $a + k\lambda \rightarrow \lambda$ defines a homeomorphism between any basis element and the entire space. But since X is Hausdorff and each basis element is closed, this implies that no neighborhood of any point is compact. For being compact it would be closed and thus contain the closure of a basis element which would then be compact even though it is homeomorphic to X which is not compact. Thus X is not locally compact.

59. The p-adic Topology on Z

Let $X = Z$ be the set of integers, and let p be a fixed prime. We define a topology τ on X by taking as a basis all sets of the form $U_\alpha(n) = \{n + \lambda p^\alpha | \lambda \in Z\}$.

1. The topology τ is generated by the metric $d(n,m) = 2^{-k}$ where k is the largest power of p which divides $|n - m|$; if $n = m$, $d(n,m) = 0$. The equivalence of these topologies follows from the relation $B_n(2^{-\alpha}) = U_{\alpha+1}(n)$.

2. Since $U_\alpha(n) = \{m | p^\alpha \text{ divides } |m - n|\} = \{m | d(m,n) \leq 2^{-\alpha}\}$, each basis set $U_\alpha(n)$ is closed. Thus X is zero dimensional. Since

no point is open, X has no isolated points; thus X is dense-in-itself, and therefore not scattered.

3. Each point is closed and not open, thus nowhere dense in X; thus since X is countable it is of the first category, hence it can be neither topologically complete, nor locally compact.

60. Relatively Prime Integer Topology

61. Prime Integer Topology

On the set $X = Z^+$ of positive integers we generate a topology τ from the basis $\mathcal{B} = \{U_a(b) | a,b \in X, \ (a,b) = 1\}$ where $U_a(b) = \{b + na \in X \ | n \in Z\}$ and a subtopology σ from the subbasis $\mathcal{P} = \{U_p(b) | p$ is prime$\}$. The topology τ will be called the relatively prime integer topology, and σ simply the prime integer topology.

1. That $\mathcal{B}$ is a basis for a topology follows directly from the observation that the intersection of any two of the arithmetic progressions in $\mathcal{B}$ is itself an arithmetic progression of the same type, or empty. In fact, if $q \in U_a(b) \cap U_c(d)$, then $U_a(b) \cap U_c(d) = U_{[a,c]}(q)$ where $[a,c]$ is the least common multiple of a and c. Clearly, $q + n[a,c] \in U_a(b) \cap U_c(d)$ for all appropriate n. Conversely if $x = b + na = d + n'c$, then since $q = b + n_0 a = d + n_0'c$ we have $x - q = (n - n_0)a = (n' - n_0')c$ so $[a,c] | x - q$ and thus $x = k[a,c] + q$ for some k, so $x \in U_{[a,c]}(q)$. So $\mathcal{B}$ is indeed a basis.

2. $U_a(b) \cap U_c(d) \neq \emptyset$ iff $b \equiv d \mod (a,c)$ for if $b - d = r(a,c)$ then there exist integers s and t such that $r(a,c) = r(as + ct) = b - d$, so $d + c(tr) = b + a(-rs)$. Thus for sufficiently large n, $d + ctr + acn \in U_a(b) \cap U_c(d)$. Conversely if for some n_0 and n_0' we have $b + n_0 a = d + n_0'c$, then $b - d = n_0'c - n_0 a = k(a,c)$, for some k, so $b \equiv d \mod (a,c)$.

3. In (X,σ), for $p \neq q$, $U_p(b)$ always intersects $U_q(d)$ since $(p,q) = 1$. So the collection $\mathcal{B}' = \{U_a(b) \in \mathcal{B} | a$ is square-free$\}$ forms a basis for (X,σ) (where an integer is called square-free if it has no repeated prime factors).

4. If $a,b \in X$, and if p is a prime greater than $a + b$, then $b \not\equiv a \mod p$ so $U_p(a) \cap U_p(b) = \emptyset$. Thus (X,σ) (and therefore also (X,τ)) is T_0, T_1, and T_2.

5. (X,τ) is not $T_{2\frac{1}{2}}$ since the closure of any open neighborhood

$U_a(b)$ contains all multiples of a, and thus the closures of any two open neighborhoods $U_a(b)$, $U_c(d)$ contain in common all multiples of $[a,c]$. To see this we observe for any k that if $(t,ka) = 1$ then $(t,a) = 1$, so $U_t(ka) \cap U_a(b) \neq \emptyset$ for $ka \equiv b \bmod (t,a)$. A similar argument may be applied to (X,σ) with similar results, so neither space can be regular or Urysohn.

6. Since both spaces are Hausdorff but not regular, they are neither locally compact, paracompact, or compact. Since $\mathcal{B}$ is countable, (X,τ) (and therefore also (X,σ)) is second countable, thus first countable. Since in second countable spaces compactness is equivalent to countable compactness, neither space can be countably compact.

7. Since in both spaces the closures of any two disjoint open sets intersect, every real-valued continuous function on (X,τ) or (X,σ) is constant. Thus both spaces are pseudocompact and connected.

8. Suppose N is an open τ-neighborhood of 1 contained in $U_2(1)$; let $1 + 2n \in N$ for some $n > 0$. Then $U = U_{2^{n+1}}(1)$ is an open subset of $U_2(1)$ whose relative complement V is open and contains $1 + 2n$ (since $V = U_2(1) - U_{2^{n+1}}(1) = \bigcup_{i=1}^{2^n-1} U_{2^{n+1}}(1 + 2i)$). Thus $U \cap N$ and $V \cap N$ separate N, so $U_2(1)$ cannot contain any open connected neighborhood of 1. Thus (X,τ) is not locally connected.

9. Suppose $U_a(b)$ is a basis element of (X,σ) with the induced topology. If A and B are open sets in (X,σ) which separate $U_a(b)$, then each contains some induced basis neighborhood: assume $N = U_{ac}(t) \subset A \cap U_a(b)$ and $M = U_{ad}(s) \subset B \cap U_a(b)$, where $(a,c) = (a,d) = 1$. Then some multiple r of cd belongs to $U_a(b)$ since $(a,cd) = 1$, and we may assume that $r \in A$. But then there is an induced basis neighborhood $U_{ae}(r) \subset A \cap U_a(b)$ where $(e,cd) = 1$; thus $(e,d) = 1$. But $U_{ae}(r) \cap U_{ad}(s) \neq \emptyset$, since $r = b + xa \equiv b + ya = s \bmod (ae,ad)$ for $(e,d) = 1$ implies $(ae,ad) = a$; so $A \cap B \neq \emptyset$, a contradiction. Thus there can be no separation of $U_a(b)$, which means that (X,σ) is locally connected since each of its basis neighborhoods is connected.

10. Returning to the space (X,τ), we see that if p is a prime $\overline{U_{p^n}(b)}$ is just $U_{p^n}(b)$ together with all nonzero multiples of p. To see this write b as $kp + \beta$ where $0 < \beta < p$ since $(p^n,b) = 1$. Let $x = mp + \gamma$ be any integer where $0 \leq \gamma < p$. Consider $U_t(x) \cap$

$U_{p^n}(b)$ where $(t,x) = 1$. This intersection is nonempty iff $mp + \gamma \equiv kp + \beta \bmod (t,p^n)$. $\gamma = 0$ iff x is a multiple of p and in this case $(t,x) = 1$ implies $(t,p^n) = 1$ so the intersection is nonempty for all t such that $(t,x) = 1$, thus $x \in \overline{U_{p^n}(b)}$. If $\gamma \neq 0$ then take $t = p^n$: $(p^n, mp + \gamma) = 1$ but $(p^n,p^n) = p^n$ so $U_{p^n}(x) \cap U_{p^n}(b) \neq \emptyset$ iff $x \equiv b \bmod p^n$. This holds iff $x \in U_{p^n}(b)$.

11. If p is a prime, $U_{p^n}(b)$ is regular open in (X,τ) except for $U_2(b)$, where $\overline{U_2(b)} = X$. For if $kp \in \overline{U_{p^n}(b)}$, then $(t,kp) = 1$ implies $(t,p^n) = 1$. Thus $U_t(kp) \cap U_{p^n}(\alpha) \neq \emptyset$ for all α where $(\alpha,p^n) = 1$; in other words, $U_t(kp)$ must contain (for each appropriate t) some element of every $\overline{U_{p^n}(\alpha)}$ for $\alpha \neq kp$. But $\overline{U_{p^n}(b)}$ may contain elements from at most two such sets: $U_{p^n}(b)$ itself, plus all multiples of p. Thus if there are more than two congruence classes (that is, if $p^n \neq 2$), then $U_t(kp)$ cannot be contained in $\overline{U_{p^n}(b)}$ for any t. Thus $\overline{U_{p^n}(b)}^\circ = U_{p^n}(b)$.

12. Since $U_a(b) = \bigcap_{i=1}^{n} U_{p_i^{\alpha_i}}(b)$ where $a = \prod_{i=1}^{n} p_i^{\alpha_i}$, the regular open sets generate the basis $\mathcal{B}$ except for sets $U_a(b)$ where $a = 2k$, $(k,2) = 1$—these being the integers that use $p^\alpha = 2$ in their prime decomposition. But each such set may be written as $U_{2k}(b) = U_{4k}(b) \cup U_{4k}(b + 2k)$. Thus the regular open sets form a subbasis, and therefore also a basis for the topology generated by $\mathcal{B}$, so (X,τ) is semiregular.

62. Double Pointed Reals

Let X be the Cartesian product of the real line with the usual topology and $\{0,1\}$ with the indiscrete topology.

1. In X, the intersection of two compact sets need not be compact. Let $A = \{[a,b] \times 0\} \cup \{(a,b) \times 1\}$, $B = \{(a,b) \times 0\} \cup \{[a,b] \times 1\}$. Since every open set containing $(a,0)$ contains $(a,1)$ both A and B are compact. But $A \cap B = (a,b) \times (0,1)$ which is not compact since (a,b) is not compact.

2. Clearly X is not T_0, T_1, or T_2; but it is T_3, T_4, and T_5.

3. Since $(a,0)$ is a limit point of every set containing $(a,1)$, X is weakly countably compact. But clearly it is neither countably compact nor pseudocompact.

4. X is arc connected, for if f is the arc which joins the points $x,y \in R$, then the function $g: [0,1] \to X$ defined by $g(t) = (f(t),1)$

for $0 \leq t < 1$, $g(1) = (f(1),0)$ is an arc joining the points $(x,1)$ and $(y,0)$. Then clearly every two points of X can be joined by an arc.

5. X is paracompact since $\{0,1\}$ is compact and the real line is paracompact. X is not fully normal since it is not T_1, but since the real line is fully normal X is fully T_4.

63. Countable Complement Extension Topology

If X is the real line, and if τ_1 is the Euclidean topology on X and τ_2 is the topology of countable complements on X, we define τ to be the smallest topology generated by $\tau_1 \cup \tau_2$.

1. A set O is open in τ iff $O = U - A$ where $U \in \tau_1$ and A is countable. So a set C is closed in τ iff $C = K \cup B$ where K is closed in τ_1 and B is countable.

2. If $O = U - A$ is open in τ, the closure in τ of O is the closure of U in τ_1, for if $K \cup B$ contains O (where B is countable, K is closed in τ_1) then $K \supset O$. So the smallest closed (in τ) set which contains O must be closed in τ_1; thus it must be $\bar{O}$.

3. So the only sets in τ which are regular open are those which were regular open in τ_1, since if $O = U - A$ is the interior of its closure $\bar{U}$, O must be U, and U must equal $\bar{U}^{\circ}$.

4. The regular open sets in τ_1 do not form a basis for this topology, for the set of irrationals is open in τ_1, yet is not the union of regular open sets. Thus X is not semiregular.

5. As an expansion of the Euclidean topology, this space is T_0, T_1, T_2, $T_{2\frac{1}{2}}$, and Urysohn.

6. Since X is T_1 but not semiregular, it cannot be T_3, T_4, or T_5. This may also be proved directly by observing that $X - Q$, the open set of irrationals, does not contain the closure of any of its open subsets, since such a closure must be identical to the usual Euclidean closure. Thus $X - Q$ cannot contain a closed neighborhood around each of its points.

7. A subset of X is compact iff it is finite, so X is neither compact nor σ-compact. But it is Lindelöf, for if $\{U_\alpha - A_\alpha\}$ is an open cover of X ($U_\alpha \in \tau_1$, A_α countable) then U_α covers X and has a countable subcover $\{U_i\}$. Then $\{U_i - A_i\}$ covers all but count-

ably many points of X, so all of X can be covered by some countable subcollection of $\{U_\alpha - A_\alpha\}$.

8. X is not first countable, for if $\{O_i\}_1^\infty = \{U_i - A_i\}_1^\infty$ were a countable collection of open neighborhoods of x and if $p \notin \bigcup_{i=1}^{\infty} A_i$, then $R - \{p\}$ is a neighborhood of x which does not contain any O_i. Since every countable set is closed, X is not separable either.

9. X is connected for if $X = O_1 \cup O_2$, where O_1 and O_2 are disjoint, nonempty and open, then they are also closed, so they must be closed in τ_1, which is impossible since (X,τ) is connected. But since the continuous image in X of the Euclidean unit interval $[0,1]$ is compact, it must be finite and therefore just a singleton. Thus X is totally pathwise disconnected, for no two points of X may be connected by a path.

64. Smirnov's Deleted Sequence Topology

Let X be the set of real numbers and let $A = \{1/n | n = 1, 2, 3, \ldots\}$. Define a topology τ on X by letting $O \in \tau$ if $O = U - B$, where $B \subset A$ and U is an open set in the Euclidean topology on R. The topology τ is sometimes called the Smirnov topology on X.

1. Choosing $B = \varnothing \subset A$, it is clear that this topology τ is finer than the usual topology on X. Therefore, X is Urysohn, as well as $T_{2\frac{1}{2}}$, T_2, T_1, and T_0.

2. X is not, however, a T_3 space, since every open set containing the closed set A intersects every open set which contains the point $0 \notin A$. Of course X also fails then to be $T_{3\frac{1}{2}}$, T_4, or T_5.

3. X is clearly not compact (since the closed subset A is not compact), but it is σ-compact since the intervals $[i,i + 1]$ for $i \neq 0$, and $[1/(i + 1),1/i]$ cover X. Since 0 does not have a compact neighborhood, X is not locally compact.

4. X is not countably paracompact, since the countable open covering by the sets $O_n = X - (A - \{1/n\})$ has no open locally finite refinement, since in every refinement every open set covering 0 must intersect infinitely many other sets of the refinement. One should note that an open set about 0 contains all of an open interval about 0 except the points $1/n$.

5. X is, however, metacompact. Suppose X is covered by open sets

$O_\alpha = U_\alpha - B_\alpha$ (where $B_\alpha \subset A$). Then $\{U_\alpha\}$ forms an open covering of the Euclidean space R, and thus has an open point finite refinement $\{V_\beta\}$. Then the collection $\{V_\beta - A\}$ is a refinement of U_α, but it covers only $X - A$. But each point of A is contained in some O_α, so we can cover each point $1/n$ by some centered open interval I_n of length less than $1/2n(n+1)$ which is contained in an O_α. These intervals will be disjoint, so the refinement of O_α consisting of $\{V_\beta - A\} \cup \{I_n\}$ covers X and is clearly point finite. Thus X is metacompact.

65. Rational Sequence Topology

Let X be the set of real numbers and for each irrational x we choose a sequence $\{x_i\}$ of rationals converging to it in the Euclidean topology. The rational sequence topology τ is then defined by declaring each rational open, and selecting the sets $U_n(x) = \{x_i\}_n^\infty \cup \{x\}$ as a basis for the irrational point x.

1. Since every Euclidean open interval contains a τ-neighborhood of each of its points, (X,τ) is an expansion of the Euclidean topology and is thus T_0, T_1, and T_2. Furthermore, each rational point and each basis set $U_n(x)$ must be closed, so (X,τ) is zero dimensional, and thus regular.

2. Any subset of X which contains a rational cannot be dense-in-itself, and any set containing an irrational could be dense-in-itself only if it contained some rational. So only the empty set is dense-in-itself, and thus X is scattered. But X is not extremally disconnected, for if $\{x_i\}$ is the rational sequence associated with the irrational point x, then $\{x_{2i}\}$ and $\{x_{2i+1}\}$ are disjoint open sets whose closures each contain x.

3. $X - Q$ is an uncountable discrete subspace, so X is not Lindelöf, and thus not second countable, though clearly it is first countable. Since Q is dense, X is separable. Clearly then, X is not metrizable.

4. X is not countably compact since the set of integers has no limit point and thus no accumulation point. Since each basis neighborhood is compact and X is T_2, X is strongly locally compact. Since X contains open points it is second category. Further, a compact set can contain only finitely many irrationals, for the irrationals in a compact set form a closed, and thus compact, discrete subset. Thus X is not σ-compact.

66. Indiscrete Rational Extension of R

67. Indiscrete Irrational Extension of R

68. Pointed Rational Extension of R

69. Pointed Irrational Extension of R

If X is the set of real numbers with the Euclidean topology τ, and if D is a dense subset of X with a dense complement, we define τ^*, the indiscrete extension of τ, to be the topology generated from τ by the addition of all sets of the form $D \cap U$ where $U \in \tau$, and τ', the pointed extension of R, to be the topology generated by all sets $\{x\} \cup (D \cap U)$ where $x \in U \in \tau$. In each case, we will be particularly interested in $D = Q$, the set of rationals, or $D = X - Q$, the set of irrationals.

1. τ' is clearly an expansion of τ since if $x \in U \in \tau$, the set $\{x\} \cup (D \cap U)$ is contained in U and open relative to τ'. Since D itself is open ($D = \{x\} \cup (D \cap X)$ if $x \in D$), τ' is also an expansion of τ^*, which is clearly an expansion of τ. Thus $\tau \subset \tau^* \subset \tau'$.

2. The set D is still dense in τ' since every neighborhood of every point of $X - D$ must contain a point of D. Therefore D is dense in τ^*.

3. Every open set O in τ' has a large closure; specifically, $\bar{O}$ is always closed in τ. For suppose every τ-neighborhood U of a point p intersects O; then $U \cap O$ is a nonempty open set in τ' which must intersect the dense set D. So if $\{p\} \cup (D \cap V)$ is a τ' neighborhood of p, it must intersect O, since $V \in \tau$, and therefore every τ-limit point of O is a τ' limit point. Thus the τ closure of O is contained in, and thus equal to the τ' closure. If $N \in \tau^*$, then $N \in \tau'$ and the τ^* closure of N is in general larger than its τ' closure. Thus the same conclusion applies to $\bar{N}$: it is closed in (X,τ).

4. Every connected subset of (X,τ') is clearly connected in (X,τ) since $\tau \subset \tau'$. The converse is also true. Since every connected subset of (X,τ) is an interval, its τ' interior equals its τ interior. Thus it suffices to show that no τ open interval S can be τ' disconnected. For if such an interval S were disconnected in τ', there would be two disjoint nonempty sets N, $M \in \tau'$ such that $S = N \cup M$. Then the τ' closure of N is τ closed, so $M =$

$(X - \bar{N}) \cap S$ is τ open; similarly, N is τ open, which is impossible since S is τ connected.

5. Since both τ^* and τ' are expansions of τ, these spaces are T_0, T_1, T_2, $T_{2\frac{1}{2}}$, and Urysohn. But in each case if O is an open set containing $X - D$, then $\bar{O}$ equals X, the τ-closure of the dense set $X - D$. So no point of D can be separated—in either topology—from $X - D$ by means of open sets. Thus neither topology is T_3, T_4, or T_5.

6. No subset of (X,τ^*) which contains a nondegenerate interval can be compact, for any such set S must contain a closed interval $[p,q]$ where $p,q \in D$. Then the sets $(-\infty,p)$, (q,∞), D, and $\{(x,q)|x > p, x \in X - D\}$ form an open covering of S with no finite subcovering. Clearly then no subset of (X,τ') which contains an interval can be compact.

7. We can prove much more: in both spaces, a set which contains an open set cannot be compact. For if the compact set C contains an open set O, it must contain the closure of O, since C is closed. But $\bar{O}$ must contain an interval so C cannot be compact.

8. Suppose f: $[0,1] \rightarrow (X,\tau^*)$ is continuous; then $f([0,1])$ is both compact and connected. But the connected sets of (X,τ^*) are precisely those of (X,τ), namely intervals, and the only intervals which can be compact are the degenerate ones consisting of one point. So f is constant, and hence (X,τ^*) (and therefore also (X,τ')) is totally pathwise disconnected. Yet both spaces are connected, since (X,τ) is. No nontrivial subset of D is connected in (X,τ) since $X - D$ is dense in (X,τ); thus clearly neither (X,τ^*) nor (X,τ') can be locally connected.

9. If $p,q \in Q$, the sets of the form (p,q) and $(p,q) \cap D$ comprise a countable base for τ^*; thus (X,τ^*) is always second countable, and therefore first countable, Lindelöf and separable. But (X,τ') is second countable iff $X - D$ is countable (as when $D = X - Q$), for then sets of the form $(a,b) \cap D$ and $\{x\} \cup ((a,b) \cap D)$ where $a,b,x \in X - D$ comprise a countable basis for τ'. On the other hand, if $X - D$ is uncountable (as when $D = Q$), (X,τ') is not even Lindelöf, for the open covering of X by sets of the form $\{x\} \cup D$, $x \in X - D$, has no countable subcover. So in this case (X,τ') is not second countable. But it is first countable since the sets $\{x\} \cup ((a,b) \cap D)$ for $a,b \in Q$ form a countable local base at x.

10. If A is compact in (X,τ'), it must be compact in (X,τ), since the identity function from (X,τ') to (X,τ) is continuous. But the only compact sets in (X,τ) which do not contain a nondegenerate interval are nowhere dense in (X,τ). So every compact subset of (X,τ'), and thus every compact subset of (X,τ^*), is nowhere dense in (X,τ). Since (X,τ) is of second category, this means that neither (X,τ^*) nor (X,τ') can be σ-compact.

70. Discrete Rational Extension of *R*

71. Discrete Irrational Extension of *R*

If X is the set of real numbers, and if D is a dense subset of the Euclidean space (X,τ) with a dense complement, we define τ^*, the discrete extension of τ to be the topology generated from τ by adding each point of D as an open set. Then any subset of D will be open in τ^*. As in the previous examples, we will be particularly interested in the cases where $D = Q$ or $D = X - Q$.

1. Since (X,τ^*) is an expansion of the Euclidean topology, it is T_0, T_1, T_2, and Urysohn.

2. No point of D is ever the limit point of a set A in (X,τ^*), while a point of $X - D$ is a limit point of A iff it is a τ-limit point of A. Thus if A and B are separated subsets of (X,τ^*), then $A - (A \cap D)$ and $B - (B \cap D)$ are separated subsets of (X,τ) and are therefore contained in disjoint open sets $N,M \in \tau$. Then since every subset of D is open, $N \cup (A \cap D)$ and $M \cup (B \cap D)$ are disjoint neighborhoods of A and B in (X,τ^*). Thus (X,τ^*) is T_5, hence regular, normal, and completely normal.

3. If D is countable (for instance, if $D = Q$) (X,τ^*) is second countable since τ has a countable basis. Thus since X is regular, it is metrizable. We can actually exhibit a metric by first enumerating D as $\{r_i\}_{i=1}^{\infty}$, and then defining $d(x,y) = \sup \{1/i | x \leq r_i \leq y\}$ whenever $x < y$, and letting $d(x,y) = 0$ if $x = y$. Then $d(r_k,x) \geq 1/k$ whenever $x \neq r_k$, so $B_\epsilon(r_k) = \{r_k\}$ whenever $\epsilon < 1/k$; thus in the metric topology, each point of D is open. Now if $x < y < z$, we have $d(x,y) \leq d(x,z)$, so each metric ball $B_\epsilon(x)$ is an interval, possibly degenerate. Clearly then, every metric ball is contained in some basis element of (X,τ^*), and every basis element of (X,τ^*) is some set $B_\epsilon(x)$. Thus (X,d) generates the topology τ^*.

4. In general, (X,τ^*) is not locally compact, for every neighborhood

N of a point $p \in X - D$ contains an infinite sequence without a limit point: simply select a point $r \in N \cap D$ and consider $\{r + 1/n\}_{n=1}^{\infty}$. For the same reason, (X,τ^*) is not countably compact.

5. If D is uncountable (for example, $D = X - Q$), (X,τ^*) cannot be separable, for any dense subset would have to contain D. Furthermore, if $X - D$ is countable, we can cover each point r_i of $X - D$ with an open interval J_i of length 2^{-i}, thus leaving uncountably many points of D uncovered. The open covering of X consisting of the intervals J_i together with each point $p \in D$ has no countable subcover. So in this case (X,τ^*) is not Lindelöf. But clearly it is first countable.

6. Since every subspace of (X,τ) is paracompact, so too is (X,τ^*). For if U is an open covering of (X,τ^*), then the collection of τ-interiors of sets in U covers some subset of X. Thus it has a locally finite refinement U^* which covers all of X except perhaps certain points of D. But these single points may be added to U^* to produce a locally finite refinement of U which covers all of X.

72. Rational Extension in the Plane

If (X,τ) is the Euclidean plane, we define a topology τ^* for X by declaring open each point in the set $D = \{(x,y)|x \in Q, y \in Q\}$, and each set of the form $\{x\} \cup (D \cap U)$ where $x \in U \in \tau$.

1. Clearly (X,τ^*) is an expansion of (X,τ), since every open set in τ contains a τ^* neighborhood of each of its points. Thus (X,τ^*) is T_0, T_1, T_2, $T_{2\frac{1}{2}}$, and Urysohn.

2. (X,τ^*) is not T_3, for if $p \notin D$, then $A = (X - D) - \{p\}$ is a closed set since $\{p\} \cup D$ is open. Yet every neighborhood $\{p\} \cup (D \cap U)$ of p intersects every neighborhood of A, since U must contain a point $q \in X - D$, and each neighborhood of q intersects $(D \cap U)$. Thus (X,τ^*) fails to be T_4 or T_5 either.

3. The open covering of X by sets $\{x\} \cup D$ where $x \in X - D$ has no countable subcover, so (X,τ^*) is neither Lindelöf or second countable. But it is separable since D is countable and dense.

4. Since D is discrete, any dense-in-itself subset must be contained in $X - D$; but every point $p \in X - D$ has a neighborhood con-

tained in $\{p\} \cup D$, so no nonempty subset can be dense-in-itself. Thus (X,τ^*) is scattered and, since it is T_1, totally disconnected.

5. But it is not totally separated, for if i is irrational, no two points $x_1 = (i,y_1)$, $x_2 = (i,y_2)$ on the vertical line $x = i$ can be separated. For if $X = A \cup B$ is a separation of X with $x_1 \in A$, $x_2 \in B$, the set $\{(i,y)|(i,y) \in A\}$ is nonempty and has a least upper bound, say (i,y_0). If $(i,y_0) \in A$, then, since A is open, there is an open set of the form $\{(i,y_0)\} \cup (D \cap U)$ contained in A; then there exists a point $(i,y') \in U$ such that $y' > y_0$. Clearly (i,y') is a limit point of A, and so is in A, which is impossible since $y' > y_0$. Similarly, (i,y_0) cannot belong to B. So there can be no such separation, and X is therefore not totally separated.

73. Telophase Topology

Let (X,τ) be the topological space formed by adding to the ordinary closed unit interval $[0,1]$ another right end point, say 1^*, with the sets $(a,1) \cup \{1^*\}$ as a local neighborhood basis.

1. (X,τ) is homeomorphic to the quotient space $[-1,1]/R$, where the equivalence classes in R are $\{-1\}$, $\{1\}$, and $\{x,-x\}$ for all $x \in (-1,1)$.
2. (X,τ) is T_1 since $[0,1]$ is, and if $a \in [0,1]$, each of the points a and 1^* have neighborhoods which do not contain the other point. But the points 1 and 1^* do not have disjoint neighborhoods, so (X,τ) is not T_2, and thus neither T_3, T_4, nor T_5.
3. Since $[0,1]$ and $[0,1) \cup \{1^*\}$ are homeomorphic as subspaces, and the subspace topology on $[0,1]$ is Euclidean, X is the union of two compact subspaces and thus compact. By the same reasoning it is arc connected.
4. $[0,1]$ and $[0,1) \cup \{1^*\}$ are compact sets with noncompact intersection.

74. Double Origin Topology

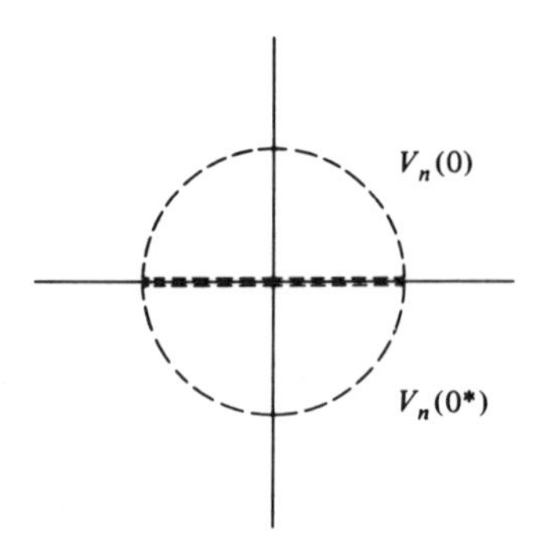

Let X consist of the set of points of the plane R^2 together with an additional point 0^*. Neighborhoods of points other than the origin 0 and the point 0^* are the usual open sets of $R^2 - 0$; as a basis of neighborhoods of 0 and 0^*, we take $V_n(0) = \{(x,y)|x^2 + y^2 < 1/n^2,\ y > 0\} \cup \{0\}$ and $V_n(0^*) = \{(x,y)|x^2 + y^2 < 1/n^2,\ y < 0\} \cup \{0^*\}$.

1. X is clearly Hausdorff, though not $T_{2\frac{1}{2}}$ since 0 and 0^* do not have disjoint closed neighborhoods: any two neighborhoods of 0 and 0^* contain a segment of the x axis in the intersection of their closures.

2. X is neither compact, paracompact, nor locally compact for if it were it would be T_3 and thus $T_{2\frac{1}{2}}$. But X is clearly second countable.

3. X is arc connected since either 0 or 0^* may be connected by an arc to any other point of X in the usual manner, except that an arc starting at 0 must be contained in the upper half-plane for a short distance, while one starting at 0^* must begin in the lower half-plane.

75. Irrational Slope Topology

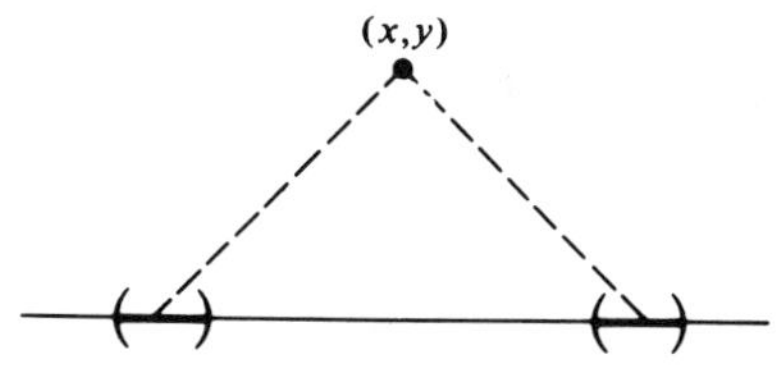

Let $X = \{(x,y) | y \geq 0,\ x,y \in Q\}$ and fix some irrational number θ. The irrational slope topology τ on X is generated by ϵ-neighborhoods of the form $N_\epsilon((x,y)) = \{(x,y)\} \cup B_\epsilon(x + y/\theta) \cup B_\epsilon(x - y/\theta)$ where $B_\epsilon(\zeta) = \{r \in Q \,|\, |r - \zeta| < \epsilon\}$, Q being the rationals on the x axis. Each $N_\epsilon((x,y))$ consists of $\{(x,y)\}$ plus two intervals on the rational x axis centered at the two irrational points $x \pm y/\theta$; the lines joining these points to (x,y) have slope $\pm\theta$.

1. (X,τ) is Hausdorff since θ is irrational, for no two points in X can lie on a line with slope θ, and if one point of X lies on a line with slope θ, no other point of X can lie on the line of slope $-\theta$ which intersects the original line at its intersection with the x axis. Thus any two distinct points in X must project (along lines with slope $\pm\theta$) onto distinct pairs of irrational points on the x axis, which have disjoint neighborhoods.

2. The closure of each basis neighborhood $N_\epsilon((x,y))$ contains the union of the four strips of slope $\pm\theta$ emanating from $B_\epsilon(x + y/\theta)$ and $B_\epsilon(x - y/\theta)$, since every point in each such ray projects to an irrational on the x axis which lies within ϵ of either $x + y/\theta$ or

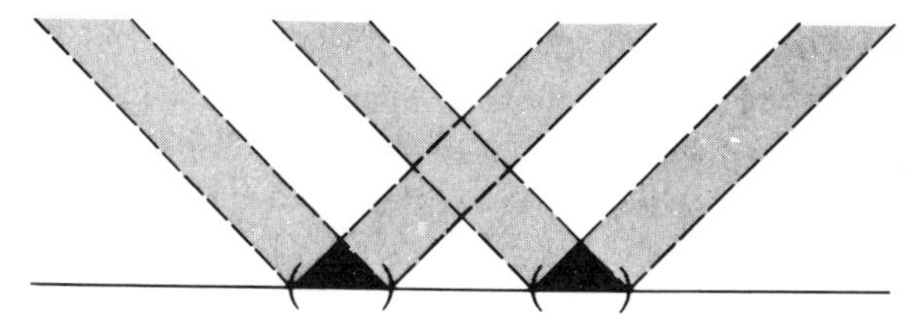

$x - y/\theta$. So, as a consequence, the closures of every two open sets must intersect. Thus (X,τ) fails to be $T_{2\frac{1}{2}}$, and consequently is neither T_3, $T_{3\frac{1}{2}}$, T_4, nor T_5.

3. Since the closure of each basis neighborhood contains open neighborhoods around each point in the diamond shaped region formed by the intersection of strips, every regular open set must contain some such diamond. Thus the regular open sets cannot form a basis for the topology, so (X,τ) is not semiregular.

4. Since the closure of any two open sets must have a nonempty intersection, (X,τ) is connected. So it is a countable connected Hausdorff space. But it cannot be path connected since if $f\colon [0,1] \to X$ is continuous, X is countable, so $\{f^{-1}(p) | p \in X\}$ is a countable collection of disjoint closed sets which covers $[0,1]$. This, however, is impossible.

5. Every real-valued continuous function f on (X,τ) is constant, for if f were not constant, $f(X)$ would contain two disjoint open sets with disjoint closures. The inverse images would then be disjoint open sets with disjoint closures, which is impossible. Thus X is pseudocompact.

6. Since X is countable and since each point of X has a countable local basis, (X,τ) is second countable and therefore has a σ-locally finite base. But it is not T_3, so it is not metrizable.

7. X is not even weakly countably compact since the sequence of integers on the x axis has no limit point.

76. Deleted Diameter Topology

77. Deleted Radius Topology

Let X be the Euclidean plane, we define the deleted diameter topology on X by taking as a subbasis for a topology σ all open discs with all of the horizontal diameters other than the center, excluded. By deleting only the horizontal radius to the right of the center we describe a subbasis for a topology τ which we call the deleted radius topology.

1. Since both σ and τ are expansions of the Euclidean topology both (X,σ) and (X,τ) are Hausdorff, completely Hausdorff, and Urysohn.

2. Since no deleted (radius or diameter) disc contains the closure of any neighborhood of its center neither (X,σ) nor (X,τ) is regular

nor semiregular. Thus neither space is locally compact. In fact, since no set with a nonempty interior is compact neither (X,σ) nor (X,τ) are σ-compact since the Baire category theorem for R^2 shows that R^2 cannot be a countable union of compact (which implies nowhere dense) sets.

3. Both (X,σ) and (X,τ) are connected since the closure of any open set in either case is its usual Euclidean closure, and thus any open and closed set must be open and closed in the Euclidean topology and thus is $\emptyset$ or X.

4. Neither (X,σ) nor (X,τ) is countably compact since both are expansions of the Euclidean topology which is not countably compact. Clearly neither space is sequentially compact.

5. That (X,σ) is neither Lindelöf nor metacompact may be shown by considering the cover consisting of the basis neighborhood at (0,0) of radius 1, the complement of the closed disc of radius 7/8 about (0,0), together with one basis neighborhood for each point not yet covered on the horizontal diameter. Since this cover has no subcover and is uncountable (X,σ) is not Lindelöf. If $\mathcal{U}$ is a fixed refinement of this cover and $x \in [-7/8, 7/8]$, let δ_x be the radius of a disc about $(x,0)$ contained in some element of $\mathcal{U}$. By the Baire category theorem for some $\epsilon > 0$ the set of $x \in I$ such that $\delta_x > \epsilon$ is not nowhere dense, so its closure contains some closed interval I'. Then the point (x,y) where $x \in I'$, $y < \epsilon$ is contained in infinitely many open sets of the refinement.

6. Similarly, (X,σ) is not countably paracompact: to show this we construct a cover by taking the basis neighborhood of radius 1 about (0,0) together with the complement of the closed disc of radius 7/8 about the origin as before. Now partition the remaining portion of the horizontal axis (including (0,0)) into a countable number of disjoint dense subsets (as in the usual construction of a nonmeasurable subset of R^1) and take for each of the remaining sets of the cover a union of basis neighborhoods of the elements of one class, where (0,0) is deleted from the class that contains it. Any neighborhood of the origin must intersect infinitely many members of any refinement of this cover, at least one for each of the countably many dense subsets.

7. We can also show that (X,τ) is not countably paracompact by choosing the same first two open sets as before, thus leaving just one interval to be covered. Consider the points $(1/n,0)$ which are

in this interval and include the point $(1/n,0)$ in the deleted disc of radius $1/(n(n+1))$ with center at $(1/n,0)$. Then in any refinement the open set containing $(0,0)$ intersects the open sets containing infinitely many of the points $(1/n,0)$.

78. Half-Disc Topology

If $P = \{(x,y)|x,y \in R,\ y > 0\}$ is the open upper half-plane with the Euclidean topology τ, and if L is the real axis, we generate a topology τ^* on $X = P \cup L$ by adding to τ all sets of the form $\{x\} \cup (P \cap U)$ where $x \in L$, and U is a Euclidean neighborhood of x in the plane.

1. Since open discs form a convenient basis for the Euclidean topology in the plane, we often think of the topology τ^* as being generated by a basis consisting of two types of neighborhoods: if $x \in P$, a basis element containing x is an open disc contained in P, whereas the basis sets around a point $y \in L$ are of the form $\{y\} \cup (P \cap D)$ where D is an open disc around y. That is, a basis set containing $y \in L$ consists of an open half-disc centered at $\{y\}$, together with $\{y\}$ itself.

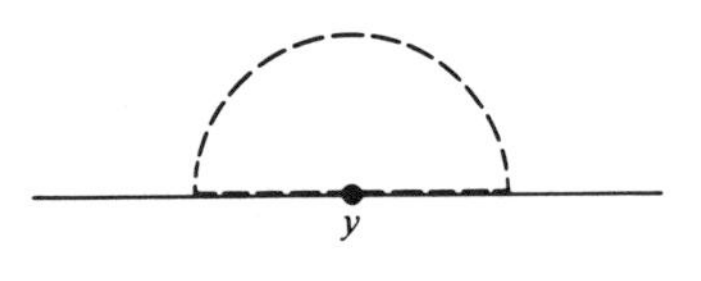

2. (X,τ^*) is an expansion of (X,τ) since every set in τ contains a τ^* neighborhood of each of its points. Thus (X,τ^*) is T_0, T_1, T_2, $T_{2\frac{1}{2}}$, and Urysohn.
3. The closure of each half-disc neighborhood of $y \in L$ includes all $z \in L$ which lie on the diameter of the disc (together with the upper circumference). So the complement of any basis neighborhood of $y \in L$ is a closed set which intersects the closure of every neighborhood of y. That is, every neighborhood of that closed set intersects every neighborhood of y; this means that (X,τ^*) is not T_3, nor, therefore, $T_{3\frac{1}{2}}$, T_4, or T_5.
4. The interior of the closure of a half-disc basis neighborhood of a point $p \in L$ contains the diameter of the disc on the x axis. Thus X is not semiregular.
5. The subspace L is discrete and uncountable, so (X,τ^*) is not second countable. Neither is it Lindelöf, for the covering by basis neighborhoods has no countable subcover. But clearly (X,τ^*) is both separable and first countable.
6. The covering of (X,τ^*) by basis elements, one for each point of X, has no point finite refinement. For consider any refinement of this

cover and let S_n denote the set of points $y \in L$ for which the element of the refinement containing y contains a basis element of radius greater than $1/n$. By the Baire category theorem for some n, $\bar{S}_n$ has a nonempty interior. Let I be an open interval contained in $\bar{S}_n$, and let $x \in I$. Then if $0 < y < 1/n$ the point (x,y) is contained in infinitely many elements of the refinement. Thus (X,τ^*) is not metacompact.

7. However X is countably metacompact. For let $\{A_i\} = \{V_j\} \cup \{U_k\}$ be a countable cover where each $V_j \subset P$, and each U_k intersects L; let $S_k = U_k \cap L$. The sets $U_k - S_k$ together with the V_j form a Euclidean open cover of the upper half-plane which may be refined to a point finite cover $\{W_\alpha\}$. Now let $T_i = S - \bigcup_{j<i} S_j$ so that the T_i are disjoint although $\bigcup_{i=1}^{\infty} T_i = L$. Define $U_k' = U_k \cap \bigcup_{s \in T_k} D_{s,k}$ where $D_{s,k}$ is a basis neighborhood of s with radius $1/k$. Since U_k' does not extend more than $1/k$ above L no point may be in more than finitely many U_k'. So $\{W_\alpha\} \cup \{U_k'\}$ covers X and is point finite.
8. If $\{T_i\}$ is a countable exhaustive collection of disjoint dense subsets of L, and if U_i is a neighborhood of T_i, then the countable cover $\{U_i\} \cup \{P\}$ has no locally finite refinement. Thus X is not countably paracompact.

79. Irregular Lattice Topology

Let X be the subset of the integral lattice points of the plane consisting of all (i,k) where $i,k > 0$, together with the points $(i,0)$ for $i \geq 0$. The lattice topology on X is determined by its basis elements: each point of the form (i,k) is itself open, each point of the form $(i,0)$ $i \neq 0$, has as a local basis sets of the form $U_n((i,0)) = \{(i,k) | k = 0 \text{ or } k \geq n\}$, while the sets $V_n = \{(i,k) | i = k = 0 \text{ or } i,k \geq n\}$ form a basis for the point $(0,0)$.

1. Clearly each open set $\{(i,k)\}$ is closed, as is each basis element $U_n((i,0))$. But the closure of V_n includes the points $(k,0)$ where $k \geq n$, since every neighborhood of these points intersects V_n. (Note that each $\bar{V}_n$ is open, though $\{\bar{V}_n\}$ does not form a local basis for the point $(0,0)$.)
2. X is a completely Hausdorff space, since it may be shown that to each pair of points $x,y \in X$, there correspond open neighborhoods O_x and O_y with disjoint closures. Since all basis elements except those around $(0,0)$ are closed, the construction of O_x and O_y is

trivial unless one point, say x, is (0,0). But in that case, if $y = (i,k)$, we need only take O_x to be V_n where $n > k$. Then $\bar{O}_x$ will be disjoint from some (closed) basis element of y.

3. X is not regular, though, since for each $n > 0$, the point (0,0) and the set $X - V_n$ do not have disjoint open neighborhoods. Thus X also fails to be completely regular, normal, and completely normal.

4. None of the sets V_n can contain a regular open neighborhood of (0,0), for the closure of any such neighborhood U must contain some point $(k,0)$ which is then an interior point of $\bar{U}$. Thus V_n cannot contain $\bar{U}^{\circ}$, so X is not semiregular.

5. Since all but one of the points of X have a local basis of sets which are both open and closed, at least one point in every pair $a,b \in X$ has this property; suppose it is a, and suppose N is an open-closed neighborhood of a disjoint from b. Then the characteristic function of N is a (continuous) Urysohn function for a and b, so X is a Urysohn space.

6. Since X is countable, and since each point of X has a countable local base, X is second countable. Furthermore, it is neither weakly countably compact, pseudocompact, nor locally compact.

7. Since each point of X is the intersection of the sets containing it which are both open and closed, each of these points is a quasi-component of X, so X is totally separated. But it is not zero dimensional and not extremally disconnected since $U = \{(1,2k) | k = 1, 2, 3, \ldots\}$ is open, yet $\bar{U} = U \cup \{(1,0)\}$ is not open.

8. X is, however, scattered since any subset which was dense-in-itself could contain no isolated points, and thus must be a subset of $S = \{(i,0) | i = 0, 1, 2, \ldots\}$. But S is discrete in the subspace topology, so it can have no nonempty dense-in-itself subsets.

80. Arens Square

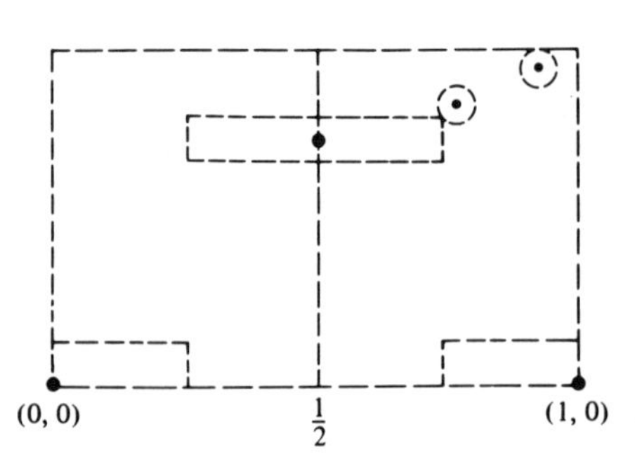

Let S be the set of rational lattice points in the interior of the unit square except those whose x-coordinate is $\frac{1}{2}$. Define X to be $S \cup \{(0,0)\} \cup \{(1,0)\} \cup \{(\frac{1}{2}, r\sqrt{2}) | r \in Q, 0 < r\sqrt{2} < 1\}$. We define a basis for a topology on X by granting to each point of S the local basis of relatively open sets which S inherits from the Euclidean topology on the unit square, and to the other points of X the following local bases:

$$U_n(0,0) = \{(0,0)\} \cup \{(x,y) | 0 < x < \tfrac{1}{4}, 0 < y < 1/n\},$$
$$U_n(1,0) = \{(1,0)\} \cup \{(x,y) | \tfrac{3}{4} < x < 1, 0 < y < 1/n\}, \text{ and}$$
$$U_n(\tfrac{1}{2}, r\sqrt{2}) = \{(x,y) | \tfrac{1}{4} < x < \tfrac{3}{4}, |y - r\sqrt{2}| < 1/n\}.$$

1. With this topology X is $T_{2\frac{1}{2}}$. This may be seen by direct consideration of cases noting that neither any point of S nor $(0,0)$ nor $(0,1)$ may have the same y coordinate as a point of the form $(\frac{1}{2}, r\sqrt{2})$.

2. X is not T_3 and hence not $T_{3\frac{1}{2}}$ since given $(0,0) \in U_n(0,0)$ there exists no open set $U_{(0,0)}$ such that $(0,0) \in U_{(0,0)} \subset \bar{U}_{(0,0)} \subset U_n(0,0)$ since $\bar{U}_{(0,0)}$ must include a point whose x coordinate is $\frac{1}{4}$ though no such point exists in $U_n(0,0)$ for any n.

3. We can show that X is not Urysohn by considering a function $f: X \to I = [0,1]$ such that $f(0,0) = 0$ and $f(1,0) = 1$. If f is continuous we note that the inverse images of the open sets $[0,\frac{1}{4})$ and $(\frac{3}{4},1]$ of I must be open and hence contain $U_n(0,0)$ and $U_m(1,0)$ for some m and n, respectively. Then if $r\sqrt{2} < \min\{1/n,1/m\}$, $f(\frac{1}{2}, r\sqrt{2})$ is not in both $[0,\frac{1}{4})$ and $(\frac{3}{4},1]$, so suppose it is not in $[0,\frac{1}{4})$: then there exists about $f(\frac{1}{2}, r\sqrt{2})$ an open interval U such that $\bar{U}$ and $[0,\frac{1}{4})$ are disjoint. But then the inverse images of $\bar{U}$ and $\overline{[0,\frac{1}{4})}$ under f would be disjoint closed sets containing open sets about $(\frac{1}{2}, r\sqrt{2})$ and $(0,0)$ respectively. But by the choice of $r\sqrt{2} < \min\{1/n,1/m\}$, these closed sets containing $U_n(0,0)$ and $U_k(\frac{1}{2}, r\sqrt{2})$ for some k cannot be disjoint.

4. X is semiregular because the basis neighborhoods are regular open sets: a straightforward check of each case reveals that $\bar{U}^\circ = U$ for each basis neighborhood U.

5. Since X is countable and since each point has a countable local basis, X is second countable. But it is neither weakly countably compact nor locally compact.

6. The components of X are each single points, and so are the quasicomponents except for the set $\{(0,0), (1,0)\}$ which is a two point quasicomponent. Thus X is totally disconnected but not totally separated.

7. X is not scattered since each basis set is dense-in-itself, nor zero

dimensional, since (0,0) cannot have a local basis consisting of open and closed sets since for sufficiently small x, the points $(x,\frac{1}{4})$ would be limit points but not interior points of each basis set.

81. Simplified Arens Square

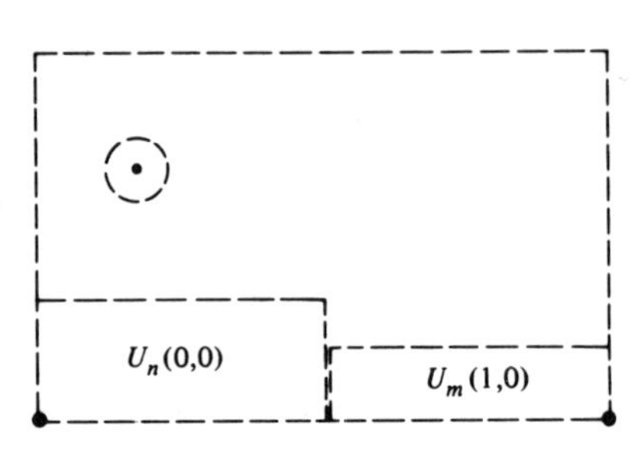

If S is the set of points in the interior of the unit square, we define X to be $S \cup \{(0,0),(1,0)\}$. Points in S will be given the Euclidean local basis neighborhoods, while $U_n(0,0) = \{(0,0)\} \cup \{(x,y)|0 < x < \frac{1}{2}, 0 < y < 1/n\}$ and $U_m(1,0) = \{(1,0)\} \cup \{(x,y)|\frac{1}{2} < x < 1, 0 < y < 1/m\}$ are the local bases for (0,0) and (1,0), respectively.

1. X may be seen to be Hausdorff by a consideration of cases. But it is not completely Hausdorff, since (0,0) and (1,0) do not have disjoint closed neighborhoods. Clearly X is semiregular, since the given basis consists of regular open sets. Thus X is not T_3, T_4, or T_5.

2. Clearly X is not sequentially compact since the induced topology on the open unit square S is the Euclidean topology.

3. X is neither locally compact nor compact for it is T_2 and not T_3.

4. X is second countable and separable since the open unit square in the Euclidean topology is, thus X is Lindelöf. But since X is not T_3, it is not paracompact and thus not countably paracompact since it is Lindelöf.

5. X is metacompact since the open unit square S is metacompact in the induced topology and the addition of a neighborhood for each of the two points (0,0) and (1,0) to a point finite refinement would not destroy its point finite character.

6. The identity map from the set X with the Euclidean topology to the given space X is continuous so X is both arc and locally arc connected.

82. Niemytzki's Tangent Disc Topology

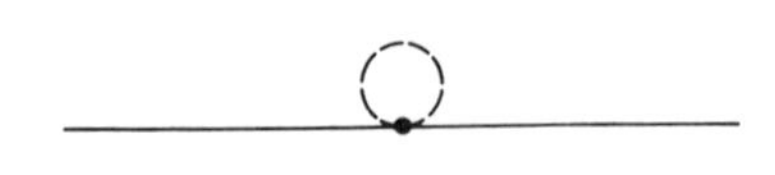

If $P = \{(x,y)|x,y \in R,\ y > 0\}$ is the open upper half-plane with the Euclidean topology τ, and if L is the real axis, we generate a topology τ^* on $X =$

$P \cup L$ by adding to τ all sets of the form $\{x\} \cup D$, where $x \in L$ and D is an open disc in P which is tangent to L at the point x.

1. (X,τ^*) is an expansion of the Euclidean subspace topology on X. For let $U \subset X$ be open in the Euclidean subspace topology, and let $x \in U$. Then if $x \in P$, there is an open neighborhood of x contained in U since for such x, the two topologies have the same local bases. If $x \in L$, then there exists a disc Δ, centered at x, and open in the entire plane, such that $\Delta \cap X \subset U$. Clearly then there is a disc $\Delta_1 \subset P$ with radius half that of Δ, which is tangent to L at x, and which is contained in Δ, and hence in U. Thus (X,τ^*) is $T_{2\frac{1}{2}}$, T_2, T_1, and T_0.

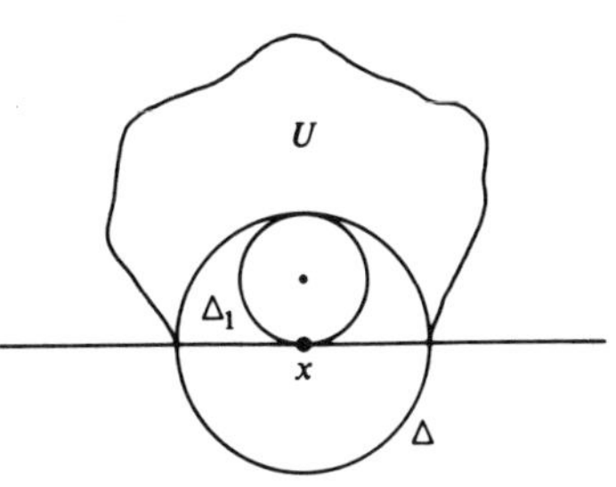

2. To show that (X,τ^*) is completely regular, we select a closed set A and a point $b \notin A$. If $b \in P$, then there is a neighborhood U of b which is contained in $X - A$, and is open in both τ^* and the Euclidean topology. So its complement, $X - U$, is closed in the Euclidean topology, and since that topology is completely regular, there is a Urysohn function for $X - U$ and b. But this function is continuous relative to τ^*, and is a Urysohn function for $A \subset X - U$ and b.

   So we consider the case where $b \in L$. There must then exist a disc D, tangent to L at b, which does not intersect A; let its radius be δ. We define a function $f: X \to [0,1]$ by requiring that $f(x) = 1$ if $x \notin D \cup \{b\}$, $f(b) = 0$, and at the point $(x,y) \in D$, $f(x,y) = [(x - b)^2 + y^2]/2\delta y$. f is continuous, since $f^{-1}([0,\alpha))$ is the open set $\{b\} \cup D_\alpha$, where D_α is the open disc of radius $\delta\alpha$ tangent to L at b and $f^{-1}((\alpha,1])$ is the open set $X - \overline{D}_\alpha$. Hence f is a Urysohn function for A and b, so (X,τ^*) is completely regular.

3. Since each basis neighborhood of each $x \in L$ contains at most one point of L, every subset of L is closed. In particular, the rationals $Q \subset L$ and the irrationals $I \subset L$ are disjoint closed sets which do not have disjoint open neighborhoods, so (X,τ^*) is not normal. For suppose $U \supset Q$ and $V \supset I$ are open sets in (X,τ^*). Then to each point $x \in V$ there corresponds a disc $D_x \subset V$ of radius r_x, tangent to L at x; let $S_n = \{x \in I | r_x > 1/n\}$. Then the collection $\{S_n\}$ together with the points of Q forms a

countable cover of the second category space (L,τ), where τ is the Euclidean topology. Thus some one of the sets S_n fails to be nowhere dense in (L,τ)—so for some integer n_0, there is an interval (a,b) in which $\{x \in L | r_x > 1/n_0\}$ is dense. Then every neighborhood of every rational in (a,b) must intersect V, so U and V cannot be disjoint.

4. X is separable since $\{(x,y) | x,y \in Q\}$ is countable and dense, but the uncountable closed subspace L is not separable, since the induced topology on L is discrete.

5. X must not be second countable since otherwise L would be second countable yet not separable. In particular, each point of L is in a basis element containing no other point of L. Since L has uncountably many points, any basis must be uncountable. Clearly then X is not Lindelöf.

6. If N is a neighborhood of the point $x \in L$, and if $D \cup \{x\}$ is a basis set where $\overline{D} \subset N$, then the circumference of $\overline{D}$ contains a sequence of points which converges to x in the Euclidean topology, but which can have no limit point in the tangent disc topology. Thus N contains a closed subset which is not countably compact, so N can be neither compact nor countably compact. Thus X is not locally compact.

7. X is not paracompact, since every paracompact Hausdorff space is normal. Furthermore, X is not countably paracompact, for suppose $\{E_i\}$ is a countable partition of L into congruent, disjoint dense subsets each of second category in L. Let $\{U_i\}$ be an open covering of X where $E_i \subset U_i$ and $U_i \cap L = E_i$; let $\{V_\alpha\}$ be a refinement of $\{U_i\}$. For each i and each n let $S_n{}^i$ be the set of points in E_i which have a tangent disc basis neighborhood in some V_α of radius $\geq 1/n$. Then for each i, at least one of the sets $S_n{}^i$ must fail to be nowhere dense in L, since $\bigcup_n S_n{}^i = E_i$ which is second category in L. So in particular for $i = 1$, there is some interval $(a,b) \subset L$ and some integer n_0 such that the refinement sets V_α contain discs of radius $\geq 1/n_0$ which are tangent to S on some dense subset F of (a,b). These discs must intersect every element of the refinement which contains a point of (a,b) and thus no point of F has a neighborhood which intersects only finitely many members of $\{V_\alpha\}$. Hence $\{U_i\}$ has no locally finite refinement, so X is not countably paracompact.

8. As in the half-disc topology, X is countably metacompact but not metacompact.

9. We can construct from X a space Y which is normal yet not completely normal since Y contains X as a subspace. To do this we merely add a point p to X, with neighborhoods $\{p\} \cup U$ where U is open in (X,τ^*), and U contains all but finitely many points of L. Then $Y = X \cup \{p\}$ is normal.

10. If we double the points of X the resulting space is only T_3 and $T_{3\frac{1}{2}}$.

83. Metrizable Tangent Disc Topology

Let S be a countable subset of the x axis in the plane. We define (X,τ) to be the subspace of the tangent disc topology consisting of $P \cup S$, where P is the open upper half-plane.

1. Since S is countable, since the rational lattice points of P are dense in X, and since X is clearly first countable, (X,τ) is second countable. Thus, since it is a subspace of a regular space, X is metrizable and thus completely normal and paracompact.

2. Let Δ be a closed disc tangent to the x axis at p. Then a sequence of points on the edge of Δ which approach p yields a countable set with no limit point. So (X,τ) is neither locally compact nor countably compact.

84. Sorgenfrey's Half-Open Square Topology

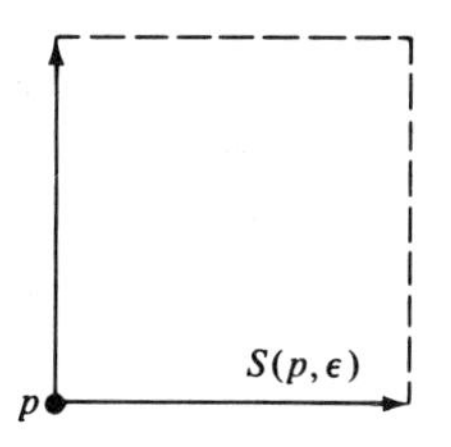

Let $S = (R,\tau)$ be the real line with the right half-open interval topology: the product space $X = S \times S$ then carries the half-open square topology in which a typical neighborhood of a point (x,y) is a rectangle (or a square) including its boundary only below the diagonal with negative slope. If $p \in X$ and if $\epsilon > 0$, we will denote the basic half-open square of side ϵ and lower left corner point p by $S(p,\epsilon)$.

1. X is completely regular since S is completely regular. (In fact, S is completely normal.)

2. Although S is completely normal, X is not even normal. This may be proved by the second category argument used previously to show that the tangent disc topology is not normal. The diagonal line $L = \{(x,y) | y = -x\}$ is closed in X and is discrete in the induced topology (since $L \cap S(p,\epsilon) = \{p\}$ whenever $p \in L$), so both $K = \{(\alpha,-\alpha) | \alpha \text{ is irrational}\}$ and its complement $L - K$

are closed in X. Then if U is a neighborhood of K, there must be some $\mu > 0$ and some subset $A \subset K$ whose Euclidean closure contains some interval I of L such that $S(p,\mu) \subset U$ whenever $p \in A$. (Otherwise, L could be written as the countable union of sets which were nowhere dense in the Euclidean topology on L.) But then no point of $(L - K) \cap I$ can have a neighborhood disjoint from U, so K and $L - K$ cannot be separated by open sets.

3. X is separable since the countable subset $D = \{(x,y)|x,y \in Q\}$ is dense in X; yet L is not separable, even though closed, for it is uncountable and discrete in the induced topology. This leads directly to an alternate proof that X is not normal. There must be 2^c continuous real-valued functions on L, but, since X is separable, there are only $2^{\aleph_0}$ continuous real-valued functions on X. So not every continuous real function on the closed subspace L can be extended to X, which means X cannot be normal. This shows, more generally, that any space with a dense set D and a closed discrete subspace S of cardinality larger than that of D cannot be normal.

4. Even though the space (S,τ) is Lindelöf, $X = S \times S$ is not Lindelöf since the closed subset L is not Lindelöf.

5. X is not metacompact since $(X - K) \cup \{S(p,1)|p \in K\}$ is an open covering of X with no point finite refinement. For, as above, any refinement of this covering must have sets which contain basic squares $B(p,\epsilon)$ for some fixed $\epsilon > 0$ and $p \in A$ where the Euclidean closure of A includes some interval of L. Clearly no such refinement can be point finite.

6. A similar argument can be used to show that X is not countably paracompact. For let $\{U_\alpha\}$ be a refinement of the countable covering $(X - (L - K)) \cup \{S(p,1)|p \in L - K\}$. Then for each point $p \in L - K$ we can select a $\delta_p > 0$ such that $S(p,\delta_p)$ is contained in some U_α; further, each U_α can contain at most one of the $S(p,\delta_p)$. If $\{U_\alpha\}$ were locally finite each $q \in K$ would have a neighborhood N_q that intersects only finitely many of the U_α, and thus only finitely many of the $S(p,\delta_p)$. Therefore with each $q \in K$ we can associate an $\epsilon_q > 0$ so that $S(q,\epsilon_q)$ is

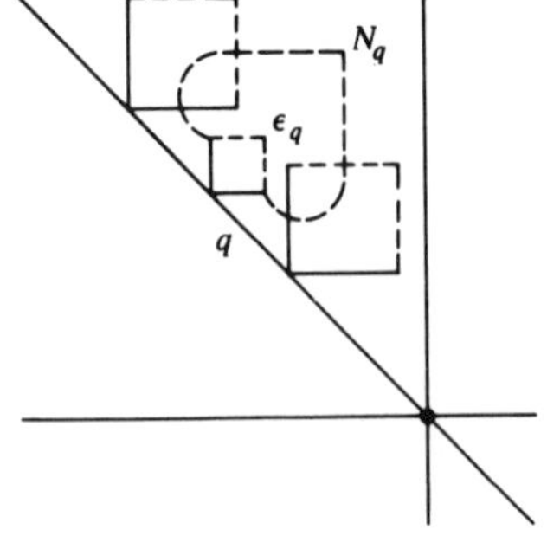

disjoint from all $S(p,\delta_p)$ where $p \in L - K$. Thus by applying the preceding category argument to the set K with its topologically complete interval topology we can guarantee the existence of an $\epsilon > 0$ and an interval $I \subset K$ such that the Euclidean closure of $\{q \in K | \epsilon_q > \epsilon\}$ contains I. So if $p \in (L - K) \cap I$, $S(p,\delta_q)$ cannot be disjoint from all $S(q,\epsilon_q)$ for $q \in K$. But this is contrary to the definition of $S(q,\epsilon_q)$, so $\{U\alpha\}$ cannot be locally finite.

85. Michael's Product Topology

If (R,τ) is the real line with the Euclidean topology and if $D = R - Q$, we define $(X,\sigma) = (R,\tau^*) \times (D,\tau')$ where τ^* is the (irrational) discrete extension of τ by the dense set D, and τ' is the induced Euclidean topology on D.

1. (D,τ') is a separable metric space, and (R,τ^*) is a completely normal paracompact space, yet the product space (X,τ) is not even normal. Consider the disjoint sets $A = Q \times D = \{(x,y) | x \in Q, y \in D\}$ and $B = \{(z,z) | z \in D\}$. A is closed since Q is closed in (R,τ^*), while B is closed since both (R,τ^*) and (D,τ') are Hausdorff. But since basic neighborhoods of points of B are vertical intervals while neighborhoods of A are rectangles, a category argument similar to that used in the tangent disc topology will show that A and B cannot have disjoint open neighborhoods in (X,σ). Specifically, any neighborhood N of B must contain, for some $\mu > 0$, intervals of length greater than μ which intersect B in a set whose Euclidean closure contains some interval. Then every neighborhood of A must intersect N.

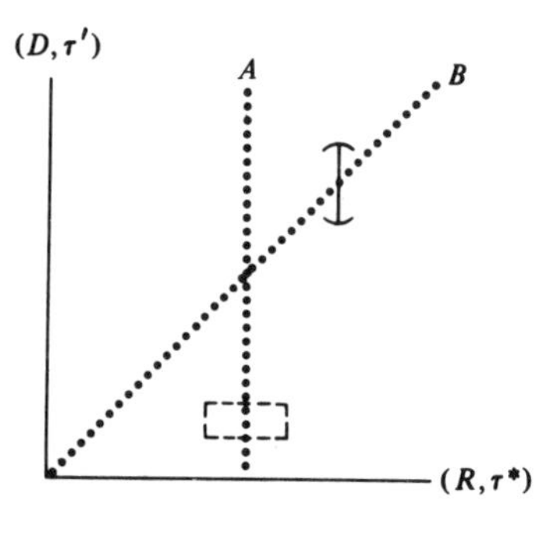

2. Since X is not normal, it is clearly not paracompact. But it is metacompact for if $\{U_\alpha\}$ is an open covering of X, the Euclidean interiors U_α° form an open covering of $\cup U_\alpha^\circ$ which has a Euclidean (and hence τ) open point finite refinement, $\{O_\alpha\}$. Now the complement $K = X - \cup U_\alpha^\circ$ is metacompact on each vertical line, so we can find on each line a point finite refinement consisting of open subsets of the line. The union of all these refinements together with $\{O_\alpha\}$ is a point finite refinement of $\{U_\alpha\}$.

86. Tychonoff Plank

87. Deleted Tychonoff Plank

If Ω is the first uncountable ordinal, and if ω is the first infinite ordinal, then the Tychonoff plank T is defined to be $[0,\Omega] \times [0,\omega]$, where both ordinal spaces $[0,\Omega]$ and $[0,\omega]$ are given the interval topology. The subspace $T_\infty = T - \{(\Omega,\omega)\}$ will be called the deleted Tychonoff plank.

1. Since ordinal space $[0,\Gamma]$ is compact and Hausdorff, so is the Tychonoff plank T. Since every compact Hausdorff space is normal, T is thus normal. (Note that even though $[0,\omega]$ and $[0,\Omega]$ are normal, we cannot conclude directly that T is normal, since normality need not be preserved under products.)

2. T fails, however, to be completely normal since the deleted plank T_∞ is not normal. For let $A = \{(\Omega,n) | 0 \leq n < \omega\}$ and $B = \{(\alpha,\omega) | 0 \leq \alpha < \Omega\}$. Then A and B are subsets of T_∞ which are closed in the subspace topology on T_∞, since their complements in T_∞ are clearly open. Now suppose $U \subset T_\infty$ is a neighborhood of A. For each point $(\Omega,n) \in A$, there is an ordinal $\alpha_n < \Omega$ such that $\{(\alpha,n) | \alpha_n < \alpha \leq \Omega\} \subset U$. Let $\bar{\alpha}$ be an upper bound for the α_n; $\bar{\alpha} < \Omega$ since Ω has uncountably many predecessors, while $\bar{\alpha}$ has only countably many. Thus the set $(\bar{\alpha},\Omega] \times [0,\omega) \subset U$. So any neighborhood of $(\bar{\alpha} + 1,\omega) \in B$ must intersect U. Thus any neighborhood V of B will intersect U, so T_∞ is not normal.

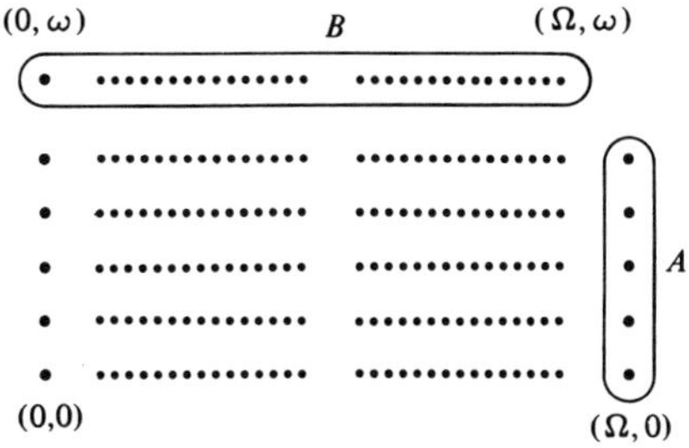

   $(0,\omega)$ B (Ω,ω) A $(0,0)$ $(\Omega,0)$

3. That T is not perfectly normal follows from the fact that T is not completely normal. It also follows directly from the observation that the closed set $\{(\Omega,\omega)\}$ is not the countable intersection of open sets. But it is the intersection of all open sets which contain it, so T cannot be first countable or separable.

4. T_∞ is not weakly countably compact, since $A = \{(\Omega,n) | 0 \leq n < \omega\}$ is an infinite set with no limit point. But it is pseudocompact, since every continuous real-valued function f on T_∞ can be extended to a continuous function $\hat{f}$ on the compact set T, and therefore both f and $\hat{f}$ are bounded. For we know that on each set $L_n =$

$\{(\alpha,n)|0 \leq \alpha \leq \Omega\}$ as well as on $L_\omega = B = \{(\alpha,\omega)|0 \leq \alpha < \Omega\}$ f is eventually constant, so for each $n \in [0,\omega]$, there exists $\gamma_n < \Omega$ such that $f((\alpha,n)) = x_n$ for all $\alpha \geq \gamma_n$. So the extension of f to $\hat{f}$: $T \to R$ given by $\hat{f}((\Omega,\omega)) = x_\omega$ will be continuous since sup $\gamma_n < \Omega$.

5. Each ordinal space is zero dimensional, scattered, but not extremally disconnected; so too are T and T_∞, for exactly the same reasons.

6. If we double the points of the Tychonoff plank the resulting space is T_3, $T_{3\frac{1}{2}}$, and T_4 only.

88. Alexandroff Plank

Let (X,τ) be the product of $[0,\Omega]$ and $[-1,1]$, each with the interval topology. If $p = (\Omega,0) \in X$, we let σ be the expansion of τ generated by adding to τ the sets of the form $U(\alpha,n) = \{p\} \cup (\alpha,\Omega] \times (0,1/n)$.

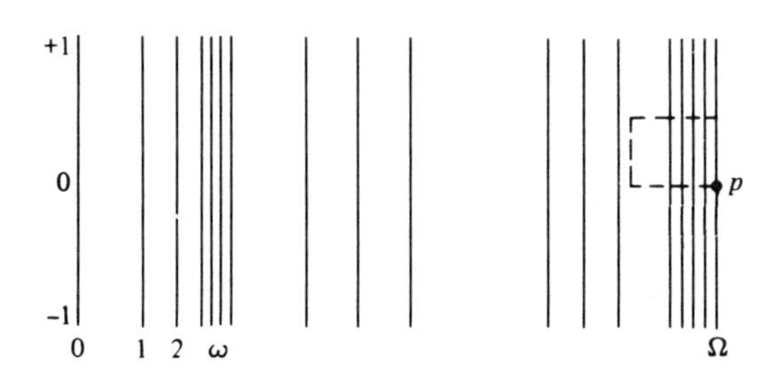

1. Since (X,τ) is regular, the expansion (X,σ) is Urysohn. But (X,σ) is not regular since $C = \{(\alpha,0)|\alpha < \Omega\}$ is a closed set not containing p, yet every neighborhood of C intersects every neighborhood of p.

2. (X,σ) is clearly semiregular since each basis rectangle in τ is regular open, as is each set $U(\alpha,n)$.

3. (X,σ) is not countably compact since the set $\{(\Omega,-1/n)|n = 2, 3, \ldots\}$ has no limit point. Neither is (X,σ) metacompact: if $\{V_\alpha\}$ is a covering of the ordinal space $[0,\Omega)$ with no point finite refinement, then the covering $\{U_\alpha\}$ of X defined by $U_1 = \{p\} \cup ([0,\Omega] \times (0,1])$, $U_2 = [0,\Omega] \times [-1,0)$, and $U_\alpha = V_\alpha \times [-1,1]$ has no point finite refinement.

89. Dieudonné Plank

Let X be $[0,\Omega] \times [0,\omega] - \{(\Omega,\omega)\}$, the points of the deleted Tychonoff plank, with the topology τ generated by declaring open each point of

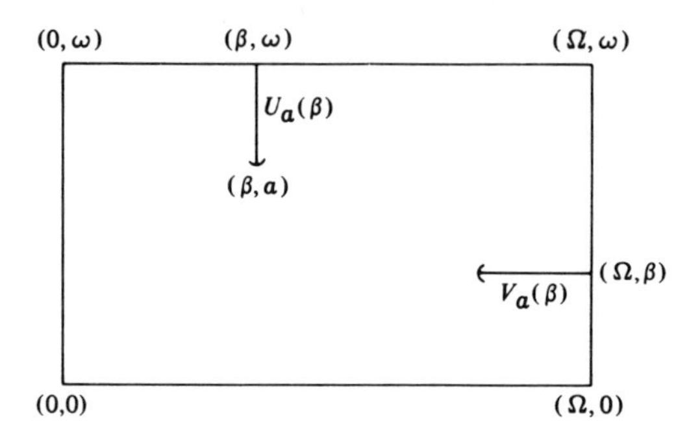

$[0,\Omega) \times [0,\omega)$, together with the sets $U_\alpha(\beta) = \{(\beta,\gamma)|\alpha < \gamma \leq \omega\}$ and $V_\alpha(\beta) = \{(\gamma,\beta)|\alpha < \gamma \leq \Omega\}$.

1. The Dieudonné topology τ on X is finer than the Tychonoff topology on $T_\infty = X$, so (X,τ) is $T_{2\frac{1}{2}}$, T_2, T_1, and T_0. Similarly, (X,τ) is not weakly countably compact.

2. Each of the open basis sets $U_\alpha(\beta)$ and $V_\alpha(\beta)$, together with each open point is also closed, so (X,τ) is completely regular and zero dimensional. But the sets $A = \{(\Omega,n)|0 \leq n < \omega\}$ and $B =$

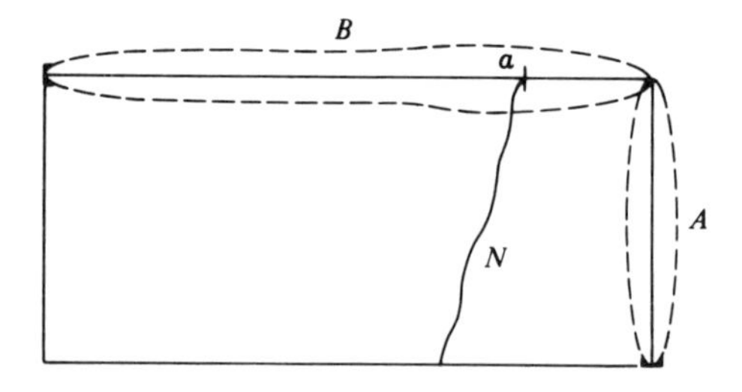

$\{(\alpha,\omega)|0 \leq \alpha < \Omega\}$ cannot have disjoint open neighborhoods since each neighborhood $N = \bigcup\{V_{\alpha_n}(n)|0 \leq n < \omega\}$ of A must intersect each basis neighborhood of each point $(\gamma,\omega) \in B$, where $\sup \alpha_n < \gamma < \Omega$. Therefore, (X,τ) is not normal.

3. (X,τ) is metacompact since any open covering of X has a refinement consisting of one basis neighborhood for each point $x \in X$. But any such refinement is point finite since an arbitrary point can be contained in at most three such basic neighborhoods.

4. The space (X,τ) is not countably paracompact since the open covering of X by the sets $U_0 = X - A$ and $U_n = V_0(n-1)$ for $n = 1, 2, 3, \ldots$ has no locally finite refinement. For if $\{W_\mu\}$

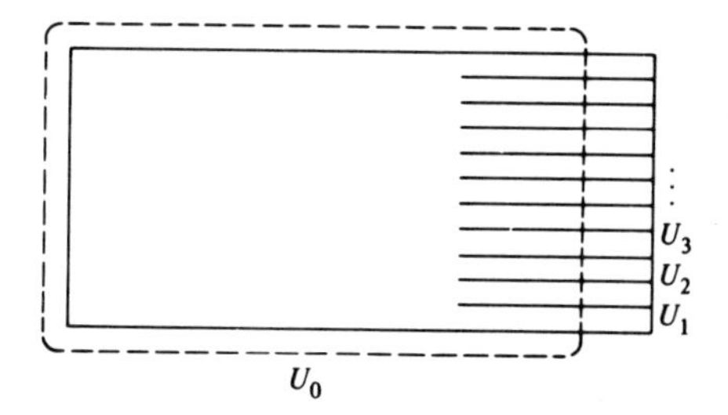

refines $\{U_n\}$, we may define for each integer n an ordinal α_n to be the least ordinal such that $V_{\alpha_n}(n)$ is contained in just one W_μ. Then if $\alpha = \sup \alpha_n < \Omega$, every neighborhood of (α,ω) will intersect infinitely many elements of $\{W_\mu\}$.

90. Tychonoff Corkscrew

91. Deleted Tychonoff Corkscrew

For each ordinal α, let A_α denote the linearly ordered set $(-0, -1, -2, \ldots, \alpha, \ldots, 2, 1, 0)$ with the order topology. Let P be the product space $A_\Omega \times A_\omega$ (where ω is the first infinite ordinal, and Ω is

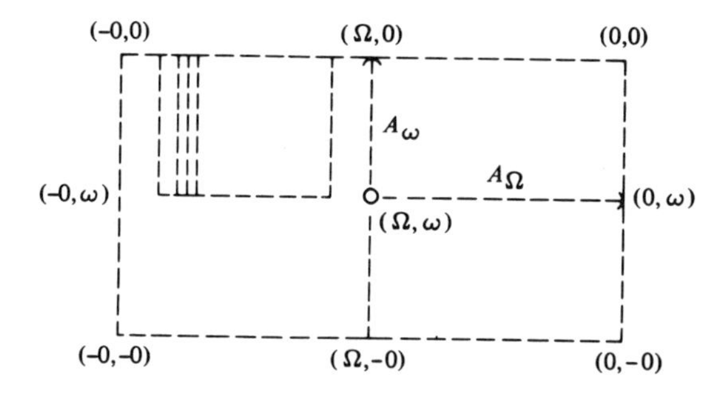

the first uncountable ordinal); let P^* be the subspace $P - \{(\Omega,\omega)\}$. Then P^* may be thought of as a rectangular lattice of points with coordinate axes A_Ω and A_ω. We then use an infinite stack of copies of P^* to form a rectangular corkscrew lattice S, spiraling in both directions, by slitting each P^* immediately below the positive A_Ω axis and then joining the fourth quadrant of one plane to the first quadrant of the one immediately below it. If $\{A_\Omega^+(i)\}_{i=-\infty}^{+\infty}$ is the indexed collection of posi-

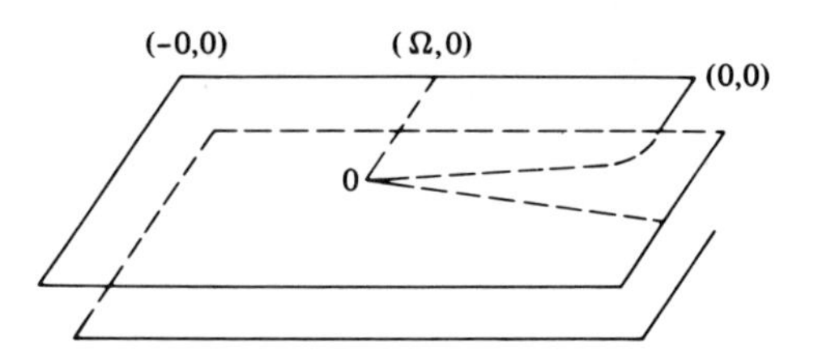

tive A_Ω axes, we will call i the level of A_Ω, and will consider, by convention, the points of S which lie above $A_\Omega{}^+(i)$ to be at level greater than i; if x is such a point, we write $L(x) > i$. To complete the infinite corkscrew X, we add to S two ideal points a^+ and a^-, to be thought of as infinity points at the top and bottom of the axis of the corkscrew; basis neighborhoods of a^+ consist of all points of X which lie above (or, for a^-,

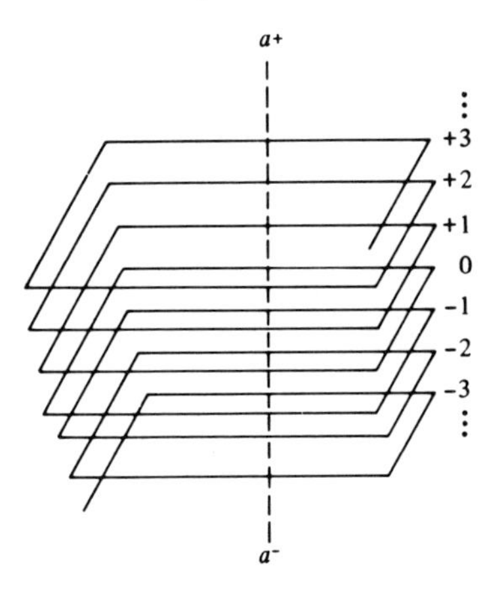

below) a certain level. The subspace $Y = X - \{a^-\}$ will be called the deleted Tychonoff corkscrew.

1. Each quadrant of the rectangular lattice P^* is homeomorphic to T_∞—the subset of the Tychonoff plank formed by deleting the point (Ω,ω). Since T_∞ is regular, the corkscrew S clearly is also.

2. To prove X regular, we thus need consider only cases involving the two ideal points a^+ and a^-. If C is a closed set in X which does not contain a^+, then $X - C$ is open and must contain all points of X above a certain level k. Then the sets $U = \{x \in X | L(x) < k + 1\}$ and $V = \{x \in X | L(x) > k + 1\}$ are disjoint open sets containing C and a^+, respectively. A similar discussion could be applied to a^-.

3. Consider now a continuous real-valued function f on X. Since the

restriction of f to each quadrant of each P^* may be extended continuously to the missing center point (Ω,ω), and since f is eventually constant on each positive and negative A_Ω axis, we see by induction that f must be constant on some set which includes at each level a deleted open interval around (Ω,ω) on the A_Ω axis. Thus there exists a sequence $\{a_i\}_{-\infty}^{\infty}$ on which f is constant, where $\lim_{i\to+\infty} a_i = a^+$ and $\lim_{i\to-\infty} a_i = a^-$. Since f is continuous, it follows that $f(a^+) = f(a^-)$, so X cannot be either Urysohn or completely regular.

4. Since X is not Urysohn, it cannot be either totally separated or zero dimensional. In fact, $\{a^+,a^-\}$ is the only quasicomponent of X containing more than one point. But X is totally disconnected since $\{a^+,a^-\}$ is discrete in the induced topology.

5. If we double the points of the corkscrew X the resulting space will be T_3 only.

6. The deleted corkscrew Y is regular since it is a subspace of X, but it is also Urysohn since every point of X except a^- can be separated from a^+ by a continuous function. However Y is still not completely regular, for, as above, the point a^+ may not be separated by a continuous function from the closed set in Y consisting of the complement of a basis neighborhood of a^+.

7. Y is totally separated since no more than one point of any given quasicomponent of X lies in Y. Further Y is not zero dimensional since a^+ had a basis of open and closed neighborhoods in Y these neighborhoods would also form a basis of open and closed neighborhoods of a^+ in X in contradiction to the fact that a^+ and a^- together form a quasicomponent of X. The basis neighborhoods of a^+ have nonopen closures so Y is not extremally disconnected. Since the Tychonoff plank is scattered so is S, and thus also Y and X.

92. Hewitt's Condensed Corkscrew

If $T = S \cup \{a^+\} \cup \{a^-\}$ is the Tychonoff corkscrew and if $[0,\Omega)$ is the set of countable ordinals, we let $A = T \times [0,\Omega)$ and define X to be the subset of A consisting of $S \times [0,\Omega)$. We think of A as an uncountable sequence of corkscrews A_λ where $\lambda \in [0,\Omega)$, and of X as the same sequence of corkscrews missing all ideal (or infinity) points. If $\Gamma\colon X \times X \to [0,\Omega)$ is a one-to-one correspondence, and if π_i $(i = 1,2)$ are the coordinate projections from $X \times X$ to X, we define a function ψ from

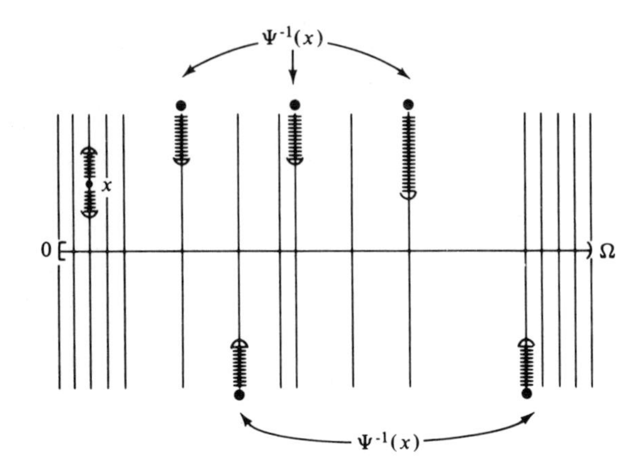

$A - X$ onto X by $\psi(a^+_\lambda) = \pi_1\Gamma^{-1}(\lambda)$ and $\psi(a^-_\lambda) = \pi_2\Gamma^{-1}(\lambda)$. Then if x and y are two distinct points of X, there exists some $\lambda \in [0,\Omega)$, namely $\lambda = \Gamma(x,y)$, such that both of the sets $\psi^{-1}(x)$ and $\psi^{-1}(y)$ intersect A_λ.

The topology on A is determined by basis neighborhoods N of each $x \in X$ with the property that $\psi^{-1}(N \cap X) \subset N$, together with A_λ-basis neighborhoods (called tails) of each point $a \in A - X$; X will inherit the subspace topology from A. To construct a typical basis neighborhood of $x \in X$, we begin with a σ-neighborhood N_0 of $x \cup \psi^{-1}(x)$ where σ is the product topology on $A = T \times [0,\Omega)$ where $[0,\Omega)$ is discrete. Then inductively, we let N_i be a σ-neighborhood of $N_{i-1} \cup \psi^{-1}(N_{i-1} \cap X)$ and $N = \cup N_i$. Clearly $\psi^{-1}(N \cap X) \subset N$.

1. X is T_1 in the induced topology since each point $x \in X$ is the intersection of all of its basis neighborhoods.

2. If $x \in X$, each basis neighborhood N of x is the union of relatively open sets $N^\lambda \subset A_\lambda \cap X$. We claim that $\bar{N}$ is simply the union of corresponding $\bar{N}^\lambda$ (where $\bar{N}^\lambda$ denotes the closure of N^λ in $A_\lambda \cap X$, and $\bar{N}$ is the closure of N in X). Suppose not; then there would be a point $y \notin \cup\bar{N}^\lambda$ every neighborhood of which intersected N. But every neighborhood M of y is a union of certain relatively open sets $M^\lambda \subset A_\lambda \cap X$, and the only way that every such set could intersect N would be for each $M \cap N$ to contain a tail in some corkscrew A_λ. But this means that the corresponding ideal point, say a^+_λ can be traced back via ψ to another tail contained in both M and N. Repeating this finitely many times produces either $y \in N$ or $x \in M$; the former contradicts the selection of y, and the latter must fail for some neighborhood M of y. Thus it is true that $\bar{N} = \cup\bar{N}^\lambda$.

3. Clearly now, since each A_λ is regular, each open basis neighborhood N contains a basis neighborhood M such that $\bar{M} \subset N$. This shows that X also is regular.

4. The function $\psi: A - X \to X$ permits a natural extension of any function f on X to a function $\hat{f}$ on A by defining $\hat{f}(x) = f(x)$ if $x \in X$ and $\hat{f}(a) = f(\psi(a))$ if $a \in A - X$. If f is continuous, so is $\hat{f}$ since for any open set U, $\hat{f}^{-1}(U) = f^{-1}(U) \cup (f \circ \psi)^{-1}(U) = f^{-1}(U) \cup \psi^{-1}(f^{-1}(U))$ where $f^{-1}(U)$ is an open subset of X. Thus $f^{-1}(U) = X \cap (\cup N_i)$ where each N_i is a σ-neighborhood in A and $\cup N_i$ is open in A. Thus
$$\begin{aligned}\hat{f}^{-1}(U) &= [X \cap (\cup N_i)] \cup [\psi^{-1}(X \cap (\cup N_i))] \\ &= [\cup(X \cap N_i)] \cup [\cup\psi^{-1}(X \cap N_i)] \\ &= \cup N_i\end{aligned}$$
since $\psi^{-1}(X \cap N_i) \subset N_{i+1}$ and $N_i \subset (X \cap N_i) \cup \psi^{-1}(X \cap N_i)$. Hence $\hat{f}^{-1}(U)$ is open in A, so $\hat{f}$ is continuous on A.

5. Every real-valued continuous function f on X is constant, for if $x,y \in X$ and if $\lambda = \Gamma(x,y)$, then $\psi(a^+{}_\lambda) = x$ and $\psi(a^-{}_\lambda) = y$. But $\hat{f}$ is continuous on A and hence on A_λ, so $\hat{f}(a^+{}_\lambda) = \hat{f}(a^-{}_\lambda)$ where $\hat{f}(a^+{}_\lambda) = f(\psi(a^+{}_\lambda)) = f(x)$ and $\hat{f}(a^-{}_\lambda) = f(\psi(a^-{}_\lambda)) = f(y)$. Thus $f(x) = f(y)$ for any two points of $x,y \in X$.

93. Thomas' Plank

94. Thomas' Corkscrew

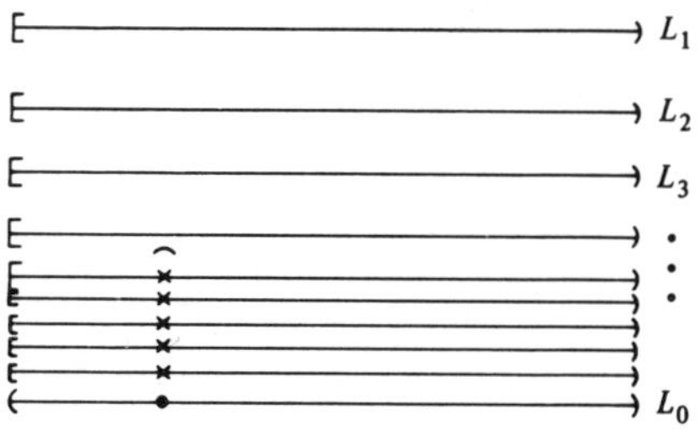

Let $X = \bigcup_{i=0}^{\infty} L_i$ be the union of lines in the plane where $L_0 = \{(x,0) | x \in (0,1)\}$, and for $i \geq 1$, $L_i = \{(x,1/i) | x \in [0,1)\}$. If $i \geq 1$, each point of L_i except for $(0,1/i)$ is open; basis neighborhoods of $(0,1/i)$ are subsets of L_i with finite complements. Similarly, the sets $U_i(x,0) = \{(x,0)\} \cup \{(x,1/n) | n > i\}$ form a basis for the points in L_0. Alternatively, if $S = L \cup \{p\}$ is the one point compactification of the unit interval $L = (0,1)$ and if $T = Z^+ \cup \{q\}$ is the one point compactification of the positive integers, then $X = S \times T - \{(p,q)\}$.

L_1

L_2

L_3

L_0

1. Every basis neighborhood of X is closed as well as open, so X is zero dimensional and completely regular (since it is clearly T_1).

2. X is not normal, since every neighborhood of the closed set $\{(0,1/n)|n = 1, 2, 3, \ldots\}$ contains all but countably many points of $\bigcup_{n=1}^{\infty} L_n$, whereas every neighborhood of the closed set L_0 contains uncountably many points of $\bigcup_{n=1}^{\infty} L_n$.

3. If f is a continuous real-valued function on X, and if $p = (0,1/n) \in L_n$, $f^{-1}(f(p)) = f^{-1}(\bigcap_{i=1}^{\infty} B_{1/i}(f(p))) = \bigcap_{i=1}^{\infty} f^{-1}(B_{1/i}(f(p)))$, a countable intersection of open sets in X. Thus $f^{-1}(f(p)) \cap L_n$ has a countable complement, so $f|_{L_n}$ is constant except for a countable set. Since f is continuous on X, $f|_{L_0}$ must also be constant except for a countable set.

4. If we use copies of Thomas' plank to build a corkscrew (as in the Tychonoff corkscrew) we can obtain a regular space which is not Urysohn, since every continuous function f will be constant except for a countable set on coordinate axes at each level of the corkscrew. Thus if p^+ and p^- are the infinity points, $f(p^+) = f(p^-)$.

95. Weak Parallel Line Topology

96. Strong Parallel Line Topology

Let A be the subset of the plane $\{(x,0)|0 < x \leq 1\}$ and let B be the subset $\{(x,1)|0 \leq x < 1\}$. X is the set $A \cup B$. We define the strong parallel line topology σ on X by taking as a basis all sets of the form

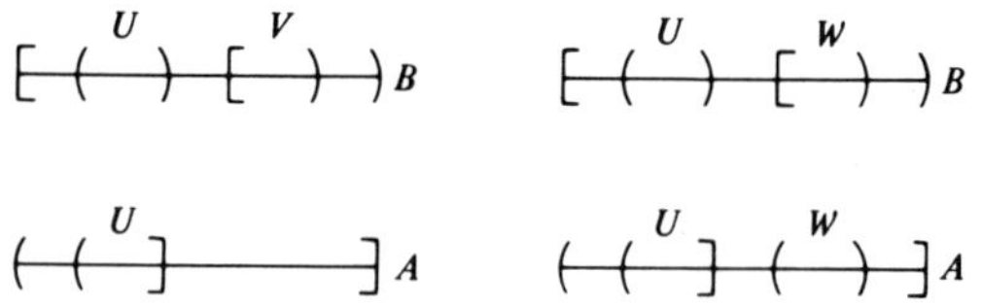

(X,σ) (X,τ)

$V = \{(x,1)|a \leq x < b\}$ and $U = \{(x,0)|a < x \leq b\} \cup \{(x,1)|a < x < b\}$, that is, left half-open intervals on B and right half-open intervals on A together with the interior of their projection onto B. The weak parallel line topology τ consists of the sets U together with sets $W = \{(x,0)|a < x < b\} \cup \{(x,1)|a \leq x < b\}$.

1. Since each W may be written as $V \cup \bigcup_{\alpha} U_\alpha$ where $V = W \cap B$ and the union is taken over all $U_\alpha \subset W$, we see that σ is a finer topology than τ.

2. If we define $(x,i) < (y,j)$ $(i,j = 0,1)$ whenever $x < y$ or $x = y$ and $i < j$ we see that for any two points p_1, p_2 of X either $p_1 < p_2$ or $p_2 < p_1$. In (X,τ), $W = \{(x,0)|a < x < b\} \cup \{(x,1)|a \leq x < b\} = \{p \in X|(a,0) < p < (b,0)\}$; since each U may be expressed similarly, (X,τ) is an order topology and is therefore completely normal. (X,σ) however is not T_3 since the closure of each V contains points on the lower line, so no open subset of B can contain a closed neighborhood.

3. If $p_1 \in A$, the sets $C = \{p \in X|p \leq p_1\}$ and $D = \{p \in X| p > p_1\}$ are both open and closed in both σ and τ. Thus $f: X \to R$ defined by $f(p) = 0$ if $p \in C$ and $f(p) = 1$ if $p \in D$ is continuous. Thus both spaces are totally separated, Urysohn, Hausdorff, and not locally connected.

4. The order which induces τ is complete, so (X,τ) is compact. For if S is a nonempty subset of X and if x is the least upper bound of the first coordinates of S, then l.u.b. S is $(x,1)$ if $(x,1) \in S$; otherwise, l.u.b. $S = (x,0)$. Since σ is finer than τ, and both are Hausdorff, σ cannot be compact.

5. Considerations entirely analogous to those given for the right half-open interval topology show that (X,σ) is neither locally compact nor σ-compact. Clearly both (X,σ) and (X,τ) are first countable, and as in the right half-open interval topology, are Lindelöf and separable, but not second countable.

6. Since no set V contains any closed neighborhood, (X,σ) is not regular, and thus is not zero dimensional since it is Hausdorff. But the closures of the sets U (or W for that matter) are just the union of the set U and a set of the form W and thus are also open. These sets form a basis for τ for each U or W is the union of all such sets contained in it. Thus (X,τ) is zero dimensional.

7. Since in either topology all the basis elements are dense-in-themselves neither space is scattered.

8. Neither (X,σ) nor (X,τ) is extremally disconnected for, as in the right half-open interval topology, a certain infinite union of disjoint basis elements has a closure which is not open.

97. Concentric Circles

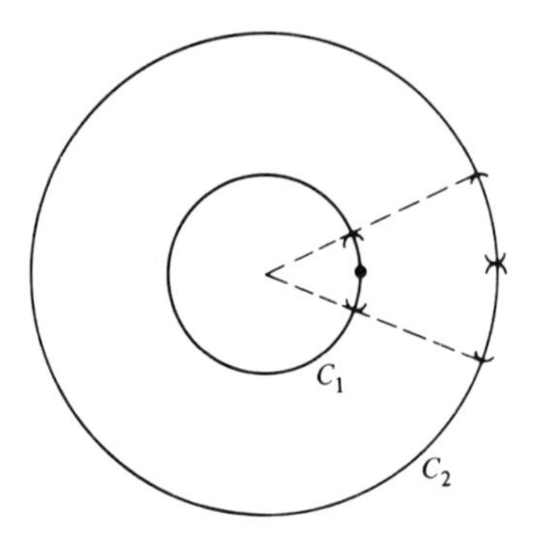

Let X consist of two concentric circles C_1, C_2 in the plane; let C_2 be the outer circle, C_1 the inner. We take as a subbasis for the topology all singleton sets of C_2, and all open intervals on C_1 each together with the radial projection of all but its midpoint on C_2.

1. That X is Hausdorff may be seen by a straightforward consideration of cases.

2. X is T_5 for if A and B are two separated sets then $A \cap C_1$ and $B \cap C_1$ are separated and thus have disjoint Euclidean neighborhoods U_A and U_B in C_1. Now if $a \in A \cap C_1$, a has a neighborhood U_a which is disjoint from B, and which may be chosen such that $U_a \cap C_1 \subset U_A$. Similarly for $b \in B \cap C_1$ we may find U_b such that $U_b \cap A = \varnothing$ and $U_b \cap C_1 \subset U_B$. Then since $A \cap C_2$ is open, $(A \cap C_2) \cup (\bigcup_{a \in A} U_a)$ and $(B \cap C_2) \cup (\bigcup_{b \in B} U_b)$ are disjoint neighborhoods of A and B, respectively.

3. X is compact. If U_α is a collection of open sets of X covering C_1, then each U_α is a union of basis neighborhoods $U_\alpha{}^\beta$. The cover of C_1 by the $U_\alpha{}^\beta$ has a finite subcover (since C_1 is compact in the Euclidean topology) so the U_α containing these finitely many $U_\alpha{}^\beta$ forms a cover of C_1 which in fact fails to cover only finitely many points of C_2. Thus any cover of X contains finitely many open sets covering all but finitely many points of C_2. One may add finitely many open sets to cover these remaining points.

4. X is sequentially compact in the Euclidean topology, so any sequence has a convergent subsequence. This subsequence still converges in the concentric circle topology τ for if its Euclidean limit point were in C_1, that point would be a τ-limit point, whereas if the Euclidean limit point were in C_2, the τ-limit point would be its projection in C_1 (unless the subsequence is constant, in which case the τ-limit point is the Euclidean limit point). Thus (X,τ) is sequentially compact.

5. X is not second countable since C_2 is an uncountable discrete subset, but X is first countable as may be seen by choosing as a local basis at the points of C_2, the points of C_2, and at the points of

C_1, nested sequences of intervals together with their deleted projections.

6. X is not separable since no countable subset is dense in C_2. Since X is compact, it is Lindelöf and thus not metrizable.
7. The components of X are C_1 and each of the points of C_2.

98. Appert Space

Let X be the set of positive integers. Let $N(n,E)$ denote the number of integers in a set $E \subset X$ which are less than or equal to n. We describe Appert's topology on X by declaring open any set which excludes the integer 1, or any set E containing 1 for which $\lim_{n\to\infty} N(n,E)/n = 1$.

1. Appert space is an expansion of countable Fort space; a set C is closed in Appert space if $1 \in C$, or if $1 \notin C$ and $\lim_{n\to\infty} N(n,C)/n = 0$.
2. Appert space is clearly Hausdorff, and in fact, completely normal. If A and B are separated sets, and if $1 \notin A \cup B$, then A and B are open. If $1 \in A$, then $\lim_{n\to\infty} N(n,B)/n = 0$ (otherwise 1 would be a limit point of B) and so B and $X - B$ are disjoint open neighborhoods of B and A respectively. If $1 \in B$, the argument is similar.
3. X is not countably compact, for the infinite set $\{2^n\}$, $n > 1$, has no limit point: no $x > 1$ can be a limit point since each such point is open, and 1 cannot be a limit point of $\{2^n\}$ since $X - \{2^n\}$ is open.
4. Since X is countable, it is σ-compact, Lindelöf, and separable.
5. X is not first countable, since the point 1 does not have a countable local basis. Suppose $\{B_n\}$ were a countable local basis at 1. Then each B_n must be infinite, so we can select an $x_n \in B_n$ such that $x_n > 10^n$; then $U = X - \{x_n\}$ does not contain any of the sets B_n, yet it is an open neighborhood of 1, for $N(n,U) > n - \log_{10} n$, and thus $\lim_{n\to\infty} N(n,U)/n = 1$.
6. X is not locally compact, since the point 1 does not have a compact neighborhood, for any neighborhood of 1 is infinite, and may be covered by a (smaller) neighborhood of 1, together with a disjoint infinite collection of open points.

7. Appert space is scattered since in T_1 spaces every dense-in-itself subset must contain an infinite number of points. But this is impossible in X since every point other than 1 is open.

8. Since any set containing 1 is closed, 1 has a local basis of sets which are both open and closed. Since all other points are discrete, X is zero dimensional.

9. X is not extremally disconnected since the set E of even integers is an open set whose closure $E \cup \{1\}$ is not open.

99. Maximal Compact Topology

Let X be the set of all lattice points (i,j) of positive integers together with two ideal points x and y. The topology τ on X is defined by declaring each lattice point to be open, and by taking as open neighborhoods of x sets of the form $X - A$ where A is any set of lattice points with at most finitely many points on each row, and as open neighborhoods of y all sets of the form $X - B$ where B is any set of lattice points selected from at most finitely many rows.

1. (X,τ) is not Hausdorff, for there are no disjoint open neighborhoods of x and y. But it is T_1 since each point is the intersection of its neighborhoods.

2. (X,τ) is compact since if $X - A$ and $X - B$ are open neighborhoods of x and y, respectively, then $X - [(X - A) \cup (X - B)] = A \cap B$ is finite.

3. Every compact subset of X is closed, for suppose that $E \subset X$ is not closed. Since each lattice point (i,j) is open, E can fail to be closed only if x or y is a limit point of, yet not in, E. Suppose $y \in \bar{E} - E$; then E must contain points from an infinite number of rows. If $A = \{(i_n,j_n)\}$ is a collection of points in E, one in each of infinitely many rows, $X - A$ together with the discrete points (i_n,j_n) forms an open covering of E but has no finite subcover. Similarly if $x \in \bar{E} - E$, then E must contain an infinite number of points from some one row; so we let B be that row and cover E by $X - B$ and singletons. Thus a set which is not closed cannot be compact.

4. Suppose $\tau^* \supset \tau$ is a compact topology for X. If the containment were strict, there would be a subset A which was closed under τ^* but not under τ. But then A would be compact under τ^* but not under τ, which is clearly impossible. Hence no topology on X which is strictly larger than τ can be compact; that is, τ is a maximal compact, non-Hausdorff topology.

5. X is not first countable since neither ideal point has a countable local basis: a Cantor-type diagonalization argument will show that no countable collection of neighborhoods of x has the property that every set of the form $X - A$ contains some set in the collection.

6. The doublet $\{x,y\}$ is a quasicomponent of X which is not a component. Thus X is totally disconnected, but not totally separated. Clearly X is scattered since in a T_1 space any dense-in-itself subset must be infinite, but all points except x and y are open.

100. Minimal Hausdorff Topology

If A is the linearly ordered set $\{1, 2, 3, \ldots, \omega, \ldots, -3, -2, -1\}$ with the interval topology, and if Z^+ is the set of positive integers with the discrete topology, we define X to be $A \times Z^+$ together with two ideal points a and $-a$. The topology τ on X is determined by the product topology on $A \times Z^+$ together with basis neighborhoods $M_n^+(a) =$

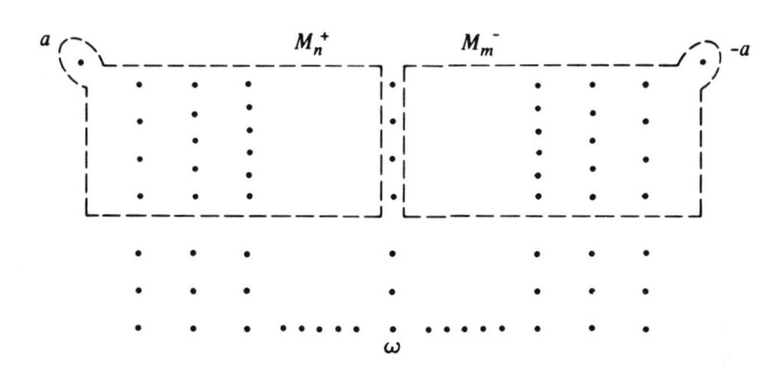

$\{a\} \cup \{(i,j) | i < \omega,\ j > n\}$ and $M_n^-(-a) = \{-a\} \cup \{(i,j) | i > \omega,$ $j > n\}$.

1. A straightforward consideration of cases shows that (X,τ) is Hausdorff, though not completely Hausdorff since for all integers n and m, $\overline{M_n^+(a)} \cap \overline{M_m^-(-a)} = \{(\omega,i) | i > \max(m,n)\} \neq \emptyset$.

2. Clearly each basis neighborhood is the interior of its closure, so (X,τ) is semiregular. But it is not regular since it is not completely Hausdorff.

3. The basis neighborhoods form an open covering of X with no finite subcovering, since the points (ω,j) are contained only in their own neighborhoods. Thus X is not compact.

4. However, X is almost compact since any collection of open sets which covers X must have a finite subcollection whose closures

cover X. This follows from the fact that the closures of any neighborhoods of a and $-a$ contain all but finitely many of the points (ω,j). A straightforward consideration of cases shows similarly that the complement of any basis neighborhood is also almost compact.

5. Now suppose $\tau^* \subset \tau$, suppose N is a basis neighborhood of τ and suppose $\{O_\alpha\}$ is a τ^*-open covering of $X - N$. It is then a τ-open covering, so there exist finitely many sets $O_1, \ldots, O_n$ the union of whose τ-closures covers $X - N$. But the τ^* closure of O_i contains the τ-closure, so $X - N$ is covered by the union of the τ^* closures of $O_1, O_2, \ldots, O_n$. In other words, $X - N$ is an almost compact subset of (X,τ^*).

6. Suppose τ^* is a proper subtopology of τ. Then there would be some basis neighborhood $N \in \tau$ for which $X - N$ would not be closed in τ^*, and so there would be a point $x \in N$ such that x belongs to the τ^* closure of $X - N$. Let $\{C_\alpha\}$ be the family of τ^*-closed neighborhoods of x, and suppose $\{X - C_\alpha\}$ covers $X - N$. Then for some $C_1 \ldots C_n$, $X - N \subset \bigcup \overline{X - C_i}$; but $\bigcup \overline{X - C_i}$ is closed, so must contain x. Since the C_i are neighborhoods of x, this is impossible, so $\{X - C_\alpha\}$ could not have covered $X - N$. Thus $\bigcap C_\alpha$ contains more than just the point x, so τ^* cannot be Hausdorff. This means that τ is a minimal Hausdorff topology for X.

101. Alexandroff Square

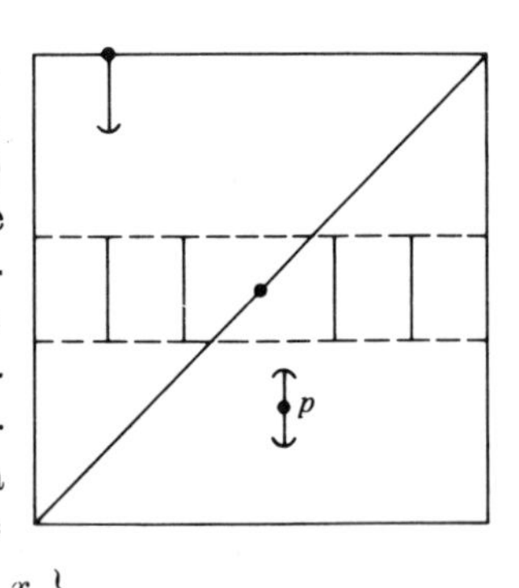

If X is the closed unit square $[0,1] \times [0,1]$, we define a topology τ by taking as a neighborhood basis of all points $p = (s,t)$ off the diagonal $\Delta = \{(x,x) | x \in [0,1]\}$ the intersection of $X - \Delta$ with an open vertical line segment centered at p: $N_\epsilon(s,t) = \{(s,y) \in X - \Delta \,\big|\, |t - y| < \epsilon\}$. Neighborhoods of points $(x,x) \in \Delta$ are the intersection with X of open horizontal strips less a finite number of vertical lines: $M_\epsilon((s,s)) = \{(x,y) \in X \,\big|\, |y - s| < \epsilon, x \neq x_0, x_1, \ldots, x_n\}$.

1. X is Hausdorff as may be seen by direct examination of cases. Consider two points (x_1,y_1), $(x_2,y_2) \in X$. If $y_1 \neq y_2$ then we may find disjoint horizontal strips, and thus disjoint open neighbor-

hoods, containing (x_1,y_1) and (x_2,y_2), respectively. If $y_1 = y_2$, then either $x_1 \neq y_1$ or $x_2 \neq y_2$; say $x_1 \neq y_1$. Then any open horizontal strip containing (x_2,y_2) less the vertical line through (x_1,y_1) and any basis neighborhood of (x_1,y_1) are disjoint neighborhoods of (x_1,y_1) and (x_2,y_2).

2. Let $\{U_\alpha\}$ be an open covering of X, and let $B = \{\alpha | U_\alpha \cap \Delta \neq \emptyset\}$. If π is the projection map of X onto the y axis, $\{\pi(U_\alpha)\}_{\alpha \in B}$ is an open covering of $[0,1]$ and thus has a finite subcovering, say $\{\pi(U_{\alpha_i})\}_{i=1}^n$. Then $\{U_{\alpha_i}\}_{i=1}^n$ covers all of X except for finitely many closed vertical line segments, and each of these may be covered by finitely many U_α. Thus (X,τ) is compact.

3. Since (X,τ) is compact and Hausdorff, it is normal. However, (X,τ) is not T_5: $A = \Delta - \{(0,0)\}$ and $B = \{(x,y) | x \geq 0,\, y = 0\}$ are clearly separated, yet they are not contained in disjoint open sets. For let U be an open set containing B. Then each $(x,0) \in B$ is contained in a vertical line of height ϵ_x contained in U. Since $[0,1]$ is uncountable, for some n the set of ϵ_x such that $\epsilon_x > 1/n$ is infinite (in fact uncountable). Then no basis neighborhood of $(1/n,1/n) \in A$ can fail to intersect U.

4. No point on the diagonal Δ can have a countable local basis so (X,τ) is not first countable. But X is sequentially compact since every sequence has a subsequence $\{(x_i,y_i)\}$ which converges in the Euclidean topology on X to a point, say (x,y). Then $\{(x_i,y_i)\}$ converges in (X,τ) to (y,y) unless all but finitely many of the x_i are equal; in this case, $\{(x_i,y_i)\}$ converges to (x,y).

5. The induced topology on each vertical line in X, as well as on the diagonal Δ, is the Euclidean topology. Since these are arc connected, so too is (X,τ) for two points in X may be joined by a path consisting of two vertical line segments and an interval along the diagonal. Thus X is connected. But clearly it is not locally connected since no nontrivial neighborhood of any point on Δ is connected.

102. Z^Z

Let Z^+ be the positive integers with the discrete topology; let $X = \prod_{i \in Z^+} Z^+_i$ be the countable Cartesian product of copies of Z^+ with the Tychonoff product topology τ.

1. If $x = \langle x_i \rangle$ and $y = \langle y_i \rangle$ are points of X, and if $E_{x,y}$ denotes $\{i \in Z^+ | x_i \neq y_i\}$, we can define a metric for X by $d(x,y) = \sum_{i \in E_{x,y}} 2^{-i}$. Since $E_{x,y} \cup E_{y,z} \supset E_{x,z}$, d is a metric for X; since $B_{2^{-i}}(a) \subset \pi_i^{-1}(a_i)$, the Cartesian product topology is finer than the metric topology. But since $\bigcap_{j=1}^{i+1} \pi_j^{-1}(a_j) \subset B_{2^{-i}}(a)$, the topologies must be equal.

2. If a subset Y of X is compact, $\pi_n(Y)$, the projection of Y onto the nth coordinate, is compact and thus finite. So no compact set in X can contain an open subset, for each open set $U \subset X$ must have some projection $\pi_n(U)$ which equals Z^+. Thus X is not locally compact.

3. X is not σ-compact either, for if $X = \bigcup Y_k$, Y_k compact, we can define $m(n,k)$ to be the greatest integer in $\pi_n(Y_k)$. Then the point whose nth coordinate is $m(n,n) + 1$ is in X, but not in any Y_k.

4. X is second countable, and thus Lindelöf, since it is the countable product of second countable spaces.

5. Since Z^+ is totally disconnected, each projection map must take a connected subset of X to a point; thus only one-point subsets of X may be connected, so X is totally disconnected. In fact, the subbasis sets $\pi_i^{-1}(j)$ are open and closed, since they are the inverse images under continuous maps of open and closed sets, so X is zero dimensional and thus totally separated.

6. No point of X is isolated so X is dense-in-itself, and thus not scattered.

7. A sequence of points $\{x_i\} \subset X$ converges to the point x iff each coordinate x_j^i eventually equals x_j (for sufficiently large i). Indeed, each coordinate in a Cauchy sequence eventually becomes constant, so X is a complete metric space.

8. In the space (X,d), the open ball $B_{2^{-j}}(a)$ is the set $\{\langle a_1, a_2, \ldots a_j, x_{j+1}, x_{j+2}, \ldots \rangle \ \exists k \ni x_{j+k} = a_{j+k}\}$, and its closure is simply the set $\{\langle a_1, a_2, \ldots, a_j, x_{j+1}, x_{j+2}, \ldots \rangle\} = \bigcap_{i=1}^{j} \pi_i^{-1}(a_i)$. Thus $\overline{B_{2^{-j}}(a)}$ is closed in (X,d) but open in (X,τ)—yet both spaces have the same topology.

9. If $e^i = \langle 1, 1, 1, \ldots, 1, 0, 0, 0, \ldots \rangle$ with i consecutive 1's,

the point $e = \langle 1, 1, 1, \ldots \rangle$ is a limit point of the set $B = \bigcup_{i=1}^{\infty} B_{2^{-2i}}(e^i)$. But $\bar{B} = \{e\} \cup \bigcup_{i=1}^{\infty} \bar{B}_{2^{-2i}}(e^i)$, so e cannot be an interior point of $\bar{B}$ since every point of $\bar{B} - \{e\}$ must contain some consecutive zeros, yet every neighborhood of e contains a point with just one zero. Hence B is an open set whose closure is not open, so X is not extremally disconnected.

10. The Baire metric on X defined by $\delta(x,y) = 1/n$ where n is the index of the first coordinate at which x and y differ has the same Cauchy sequences as the metric d. Thus it yields the same topology.

103. Uncountable Products of Z^+

If Z^+ is the discrete space of positive integers, we let $X_\lambda = \prod_{\alpha \in A} Z^+_\alpha$, where λ is the cardinality of A. We assume that $\lambda > \aleph_0$—that is, that A is uncountable.

1. X_λ is clearly Hausdorff and completely regular since these properties are preserved by arbitrary products.

2. X_λ is neither first nor second countable. Assume $\{B_i\}$ is a countable local base at the point $p \in X_\lambda$. For each i, $\pi_\alpha(B_i) = Z^+$ for all but finitely many α; since there are uncountably many α, we can select one, say α_0, such that $\pi_{\alpha_0}(B_i) = Z^+$ for all i. Then $\pi_{\alpha_0}^{-1}(p_{\alpha_0}) = \{y \in X_\lambda | y_{\alpha_0} = p_{\alpha_0}\}$ is an open neighborhood of p which contains no B_i, so $\{B_i\}$ is not a local base at p.

3. If $\lambda \leq 2^{\aleph_0}$, X_λ is separable. For in this case there exists a bijection ϕ of A onto a proper subset of the unit interval I. Let $J_1, J_2, \ldots, J_k$ be any finite pairwise disjoint collection of closed subintervals of I with rational endpoints, and $n_1, n_2, \ldots, n_k$ be a finite subset of Z^+. Let $p(J_1, \ldots, J_k, n_1, \ldots, n_k)$ be the point $\langle p_\alpha \rangle \in X_\lambda$ where $p_\alpha = n_i$ if $\phi(\alpha) \in J_i$, and $p_\alpha = 0$ otherwise.

   The set of all such points $p(J_1, J_2, \ldots, J_k, n_1, n_2, \ldots, n_k)$ is clearly countable; it is also dense, for given any open set in X_λ of the form $\bigcap_{j=1}^{k} \pi_{\alpha_j}^{-1}(U_{\alpha_j})$, where U_{α_j} is any open set (that is, any set) in $Z^+_{\alpha_j}$, we can find a pairwise disjoint collection $J_1, \ldots, J_k$ such that $\phi(\alpha_j) \in J_j$, and for each j, an n_j such that $n_j \in U_{\alpha_j}$. Then $p(J_1, \ldots, J_k, n_1, \ldots, n_k) \in \bigcap_{j=1}^{k} \pi_j^{-1}(U_{\alpha_j})$, since whenever $\phi(\alpha_j) \in J_j$, $p_\alpha = n_j \in U_{\alpha_j}$.

4. Conversely, if $\lambda > 2^{\aleph_0}$, X_λ is not separable. For suppose D were a countable dense subset of X_λ; then for each $\alpha,\beta \in A$, the sets $D \cap \pi_\alpha^{-1}(1)$ and $D \cap \pi_\beta^{-1}(1)$ are distinct, since each basis neighborhood $\pi_\alpha^{-1}(1)$ and $\pi_\beta^{-1}(1)$ is both open and closed. Thus the map $\Phi\colon A \to P(D)$ given by $\Phi(\alpha) = D \cap \pi_\alpha^{-1}(1)$ is injective, so $\lambda = \text{card } A \leq \text{card } P(D) = 2^{\aleph_0}$.

5. X_λ is neither compact, σ-compact, nor locally compact since Z^Z is none of these and they are all preserved by open continuous maps. Also like Z^Z, X_λ is zero dimensional—since each subbasis set $\pi_\alpha^{-1}(p_\alpha)$ is both open and closed—but not scattered.

6. For $i = 0$ and 1, let $P_i \subset X_\lambda$ be the collection of all points with the property that each integer except i appears at most once as a coordinate. Then since A is uncountable, $P_1 \cap P_0 = \emptyset$ and since $X_\lambda - P_i = \bigcup_{\substack{\alpha\neq\beta\\ n\neq i}} (\pi_\alpha^{-1}(n) \cap \pi_\beta^{-1}(n))$, P_i is closed. We will show that any two neighborhoods U and V of P_0 and P_1 intersect, so X_λ cannot be normal.

   Each finite subset $F \subset A$ determines a basis neighborhood $F(x)$ of any point $x \in X_\lambda$ by $F(x) = \bigcap_{\alpha\in F} \pi_\alpha^{-1}(x_\alpha)$. We define inductively a nested increasing sequence $F_n = \{\alpha_j\}_{j=1}^n$ of finite subsets of A together with an associated sequence of points $x^n \in P_0$. Let $x_\alpha^0 = 0$ for all $\alpha \in A$, and suppose that x^n and F_{n-1} are given; select $F_n \supset F_{n-1}$ by requiring that $F_n(x^n) \subset U$, and then select $x^{n+1} \in P_0$ so that $x_{\alpha_j}^{n+1} = j$ whenever $\alpha_j \in F_n$, and $x_\alpha^{n+1} = 0$ otherwise. Now let $y \in P_1$ be defined by $y_{\alpha_j} = j$ whenever $\alpha_j \in \bigcup F_n$, and $y_\alpha = 1$ otherwise. Then there is a finite set $G \subset A$ such that $G(y) \subset V$. Then for some integer m, $G \cap \bigcup_{n=0}^{\infty} F_n = G \cap F_m$, so we may define a point $z \in X_\lambda$ by $z_{\alpha_k} = k$ whenever $\alpha_k \in F_m$, $z_{\alpha_k} = 0$ whenever $\alpha_k \in F_{m+1} - F_m$, and $z_\alpha = 1$ otherwise. Then $z_\alpha = y_\alpha$ if $\alpha \in G \cap F_m$, and otherwise if $\alpha \in G$, $z_\alpha = y_\alpha = 1$; thus $z \in G(y) \subset V$. Furthermore, $z_{\alpha_k} = k = x_{\alpha_k}^{m+1}$ if $\alpha_k \in F_m$, and $z_{\alpha_k} = 0 = x_{\alpha_k}^{m+1}$ if $\alpha_k \in F_{m+1} - F_m$; thus $z \in \bigcap_{\alpha\in F_{m+1}} \pi_\alpha^{-1}(x_\alpha^{m+1}) = F_{m+1}(x^{m+1}) \subset U$. So $z \in U \cap V$, which was to be proved.

104. Baire Metric on R^ω

If $X = R^\omega$ is the set $\prod_{i=1}^{\infty} R_i$, where each R_i is a copy of the Euclidean real line, we define the Baire metric on X by $d(\langle x_i\rangle,\langle y_i\rangle) = 1/i$ where i is the first coordinate where x and y differ.

1. A Baire basis set $B_{1/i}(x)$ consists of all $y \in X$ whose first i coordinates agree with those of x. Clearly no basis set of the Tychonoff product topology on R^ω can be contained in $B_{1/i}(x)$, but $B_{1/i}(x)$ is contained in the Tychonoff basis set $\bigcap_{j=1}^{i} \pi_j^{-1}(U_j)$ where $x_j \in U_j$ for all $j \leq i$. Thus the Baire topology is strictly finer than the Tychonoff topology on R^ω.

2. If $\{x_n\}$ is a Cauchy sequence in X, and if $x_{i,n}$ is the ith coordinate of the term x_n, then for each i the sequence $x_{i,1}, x_{i,2}, x_{i,3}, \ldots$ is eventually constant, say x^i. Then clearly $\{x_n\}$ converges to $x = \langle x^i \rangle$. Thus X is complete.

3. The induced topology on each coordinate axis R_i is discrete. In fact, since $B_{1/i}(x) = \bigcap_{j=1}^{i} \pi_j^{-1}(x_j)$, the topology generated on R^ω by the Baire metric is precisely the Tychonoff product topology where the factors R_i are assumed discrete. But clearly X is not discrete.

4. Since the projection of every compact subset of X onto each discrete factor space R_i is compact and therefore finite, a compact subset of X cannot contain an open set. Hence every compact set is nowhere dense. So by the Baire category theorem, X cannot be σ-compact. Similarly, X cannot be locally compact.

5. Since the discrete topology on each factor space R_i is non-separable, X cannot be separable.

6. Since the subbasis sets $\pi_i^{-1}(x_i)$ are the inverse images under continuous maps of sets which are both open and closed, they, and therefore the corresponding basis sets, are both open and closed. Thus X is zero dimensional. Since no point is open, X is dense-in-itself and therefore not scattered; since X is metrizable and not discrete, it is not extremally disconnected.

105. I^I

Let I^I be the uncountable Cartesian product of the closed unit interval $I = [0,1]$: $I^I = \prod_{i \in I} I_i$.

1. I^I is compact and Hausdorff since I is compact and Hausdorff. Thus I^I is normal. Similarly since I is connected, so is I^I.

2. If $A = \{1/n \in I | n \in Z^+\}$, I^I contains $Y = \prod_{i \in I} A_i$; the subspace topology on Y is homeomorphic to $\prod_{i \in I} Z^+_i$ since the induced topology on A is homeomorphic to the discrete topology on Z^+. Thus Y is a subspace of I^I which is not normal, so I^I cannot be completely normal.

3. Suppose $\{B_n\}$ is a countable local basis at a point $y \in I^I$. Since for each n, $\pi_\alpha(B_n) = I$ for all but finitely many α, and since I is uncountable, there must be an α_0 such that $\pi_{\alpha_0}(B_n) = I$ for all n. So if U is an open neighborhood of y_{α_0}, $U \neq I$, $\pi_{\alpha_0}^{-1}(U)$ is an open neighborhood of y which contains no B_n. Thus I^I is not first countable.

4. Points of I^I are functions from I to I, and a sequence of points α_k converges in I^I to α iff the functions α_k converge pointwise to α—that is, iff for each $x \in I$, $\alpha_k(x)$ converges in I to $\alpha(x)$. This equivalence follows directly from the definition of the product topology on I^I, for open neighborhoods in I^I restrict only finitely many coordinates at a time, and this is precisely pointwise convergence.

5. I^I is not sequentially compact since the sequence of functions $\alpha_n \in I^I$ defined by $\alpha_n(x) =$ the nth digit in the binary expansion of x has no convergent subsequence. For suppose $\{\alpha_{n_k}\}$ is a subsequence which converges to a point $\alpha \in I^I$. Then for each $x \in I$, $\alpha_{n_k}(x)$ converges in I to $\alpha(x)$. Let $p \in I$ have the property that $\alpha_{n_k}(p) = 0$ or 1 according to whether k is odd or even. Then the sequence $\{\alpha_{n_k}(p)\}$ is 0, 1, 0, 1, . . . , which cannot converge.

106. $[0,\Omega) \times I^I$

Let X be the product of $[0,\Omega)$ with the interval topology and I^I with the Cartesian product topology.

1. X is Hausdorff and completely regular since both factor spaces are.

2. Since $[0,\Omega)$ is countably compact and I^I is compact, X is countably compact. However X is neither compact nor sequentially compact since these are preserved by projections and $[0,\Omega)$ fails to be compact while I^I fails to be sequentially compact.

3. Since $[0,\Omega)$ is neither separable, Lindelöf, nor σ-compact, X also cannot satisfy these conditions. Similarly, since I^I is not first countable, neither is X. But since both I^I and $[0,\Omega)$ are locally compact, so is X.

107. Helly Space

The subspace X of I^I consisting of all nondecreasing functions is called Helly space; it carries the induced topology.

1. If $f \in I^I - X$, it is not nondecreasing. Thus there are points $x,y \in I$ such that $x < y$ but $f(x) > f(y)$. If $\epsilon < \frac{1}{2}[f(x) - f(y)]$, and if $U_x = B_\epsilon(f(x))$ and $U_y = B_\epsilon(f(y))$, then $\pi_x^{-1}(U_x) \cap \pi_y^{-1}(U_y)$ is a neighborhood of f disjoint from X. Thus X is a closed subspace of the compact set I^I, so X is a compact Hausdorff space.

2. Each function $f \in X$ has at most countably many points of discontinuity. Let A_f be these points of discontinuity together with the rationals in the unit interval I; then A_f is a countable set. We claim that the set of all finite intersections of $\pi_a^{-1}(B_{1/j}(f(a)))$ for $a \in A_f$ and $j = 1, 2, 3, \ldots$ is a countable local basis for f. To verify this we need only check that every subbasis neighborhood $\pi_y^{-1}(B_\epsilon(f(y)))$ contains an element of our countable family. If y is a point of discontinuity of f, we merely select j so that $1/j < \epsilon$, and take $\pi_y^{-1}(B_{1/j}(f(y)))$. If not, then f is continuous at y, so there is a δ such that $f(x) \in B_\epsilon(f(y))$ whenever $|x - y| < \delta$. So if a is a rational in the interval $(y,y + \delta)$ and j so large that $B_{1/j}(f(a)) \subset B_\epsilon(f(y))$, and similarly if b is a rational in $(y - \delta,y)$ and $B_{1/k}(f(b)) \subset B_\epsilon(f(y))$, then every function of X in $\pi_a^{-1}(B_{1/j}(f(a))) \cap \pi_b^{-1}(B_{1/k}(f(b)))$ must be in $\pi_y^{-1}(B_\epsilon(f(y)))$. Thus X is first countable.

3. Let A_i be the set of diadic rationals in I with denominator 2^i; then $\{A_i\}$ is an increasing nested sequence of finite subsets of I whose union is dense in I. For each i, let Y_i be the set of continuous piecewise linear functions in X which take on rational values at each point of A_i and which are linear between these points. Each Y_i is countable, so $Y = \bigcup Y_i$ is also. Furthermore, Y is dense in X, so X is separable.

4. The collection of functions f_x defined by $f_x(t) = 0$ if $t < x$, $f_x(x) = \frac{1}{2}$, and $f_x(t) = 1$ if $t > x$ is an uncountable subset of X

which is discrete in the induced topology. Thus, this subspace is not second countable, and so therefore neither is X; furthermore, since X is separable, it cannot be metrizable.

108. $C[0,1]$

The space of real-valued continuous functions on the unit interval I will be denoted by $C[0,1]$; it is a metric space under the sup norm distance: $d(f,g) = \sup_{t \in I} |f(t) - g(t)|$.

1. $C[0,1]$ is a complete metric space, since each Cauchy sequence in $C[0,1]$ is a uniformly convergent sequence of continuous functions, which must have a continuous function as a limit.

2. $C[0,1]$ is separable since the polynomials with rational coefficients form a countable dense subset.

3. Every open ball $B_\epsilon(f)$ contains a sequence $\{f_i\}$ such that $d(f_i,f_j) = \epsilon$ (for $i \neq j$) and which therefore has no convergent subsequence. To see this we define $\phi_n \in C[0,1]$ to be $-\epsilon/2$ on $[0,1/2 - \epsilon/2n]$, $+\epsilon/2$ on $[1/2 + \epsilon/2n, 1]$ and linear in between. Then $f_i = f + \phi_i \in B_\epsilon(f)$ and satisfies $d(f_i,f_j) = \epsilon$. Thus $C[0,1]$ has no compact neighborhoods, so is neither locally compact nor, since it is of second category, σ-compact.

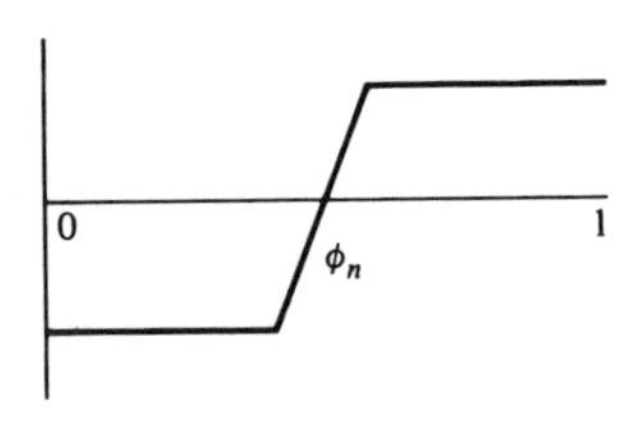

4. $C[0,1]$ is both arc connected and locally arc connected since each ball $B_\epsilon(f)$ is convex. That is, the function $\phi\colon [0,1] \to C[0,1]$ defined by $\phi(t) = tg + (1 - t)h$ is a path joining g to h; if g and h lie in $B_\epsilon(f)$, so does each $\phi(t)$.

109. Box Product Topology on $\mathbf{R}^\omega$

If $X = R^\omega$ is the set $\prod_{i=1}^{\infty} R_i$, where each R_i is the Euclidean real line, we generate the box product topology τ from basis sets of the form $\prod_{i=1}^{\infty} U_i$ where each U_i is open in R_i.

1. Clearly each basis set of the Tychonoff topology is open in the box topology, so τ is strictly finer than the Tychonoff product topology on X. Thus (X,τ) is Hausdorff.

2. (X,τ) is completely regular. Let U be an open set and let $x = \langle x_i \rangle \in \prod_{i=1}^{\infty} U_i \subset U$, where each U_i contains an open interval $(x_i - \epsilon_i,\ x_i + \epsilon_i)$. Define $f(x) = 0$ and for $y \in X - \{x\}$ define $f(y) = \inf\{r \in (0,1] |$ for some i, $y_i \not\subset (x_i - r\epsilon_i,\ x_i + r\epsilon_i)\}$. Then f is a Urysohn function for x and $X - U$.
3. X is not connected since the set of bounded sequences and the set of unbounded sequences are both open.
4. Since (X,τ) is an expansion of the Tychonoff topology on R^ω, which is homeomorphic to Hilbert space, any compact subset of X must be compact in Hilbert space. Thus, since Hilbert space is not σ-compact, neither is (X,τ).
5. (X,τ) is not locally compact since the origin, for instance, has no compact neighborhoods. For let U be an open set containing the origin, and let $x = \langle x_i \rangle \in U$. Then the sequence $x_n = \langle x_i/n \rangle$, which ought to converge to the origin, does not do so, because $\prod_{i=1}^{\infty} (-x_i/2i,\ x_i/2i)$—a neighborhood of the origin—excludes every x_n.
6. Suppose $\{U_i\}$ is a countable local basis at the point $x \in X$; let $U_i = \prod_{j=1}^{\infty} U_{ij}$. Then for each i, let V_i be a proper subset of U_{ii} containing x_i; then $\prod_{i=1}^{\infty} V_i$ is a neighborhood of $x = \langle x_i \rangle$ which does not contain any of the sets $\{U_i\}$. Thus $\{U_i\}$ cannot be a basis at x, so X is not first countable.
7. Neither is X separable, for if, for each i, U_{ij} is the open interval $(j, j+1)$ in R_i, then the collection of all sets $A_{\{j_i\}} = \prod_{i=1}^{\infty} U_{i,j_i}$, where $\{j_i\}$ is an infinite subset of the integers, forms an uncountable disjoint collection of open sets in X. This shows in fact that X does not satisfy the countable chain condition.

110. Stone-Čech Compactification

Let (X,τ) be a completely regular space, let I be the closed unit interval $[0,1] \subset R$, and let $C(X,I)$ be the collection of all continuous functions from X to I. Let $I^{C(X,I)} = \prod_{\lambda \in C(X,I)} I_\lambda$ where I_λ is a copy of I indexed by $\lambda \in C(X,I)$. We denote by $\langle t_\lambda \rangle$ the element of $I^{C(X,I)}$ whose λth coordinate is t_λ. Then if $h_X\colon X \to I^{C(X,I)}$ is defined by $h_X(x) = \langle \lambda(x)_\lambda \rangle$, the image of h_X, $h_X(X)$, is a subset of the compact Hausdorff space $I^{C(X,I)}$, so its closure $\beta X = \overline{h_X(X)}$ is a compact Hausdorff space known as the Stone-Čech compactification of (X,τ).

1. h_X is a homeomorphism of X onto $h_X(X)$, a dense subset of βX. Since $p_\lambda \circ h_X = \lambda$ for each $\lambda \in C(X,I)$ (where p_λ is the projection of ΠI_λ onto I_λ), h_X is continuous. It is also injective since if $x,y \in X$, $x \neq y$, there is a continuous function $\lambda\colon X \to I$ such that $\lambda(x) = 0$ and $\lambda(y) = 1$. Thus $h_X(x) \neq h_X(y)$. In fact h_X is injective iff X is Urysohn. Finally, we can show that h_X is an open mapping by selecting any open set $U \subset X$, and a point $x \in U$. Since (X,τ) is completely regular, there exists $\lambda\colon X \to I$ such that $\lambda(x) = 0$ while $\lambda \equiv 1$ on the closed set $X - U$. Then $h_X(X) \cap p_\lambda^{-1}((-\infty,1))$ is an open subset of $h_X(U)$ which contains $h_X(x)$. So $h_X(U)$ is open.

2. Every continuous function f from X to a compact Hausdorff space Y has a unique continuous extension to βX; that is, for each continuous $f\colon X \to Y$ there exists a continuous function $\hat{f}\colon \beta X \to Y$ such that $\hat{f} \circ h_X = f$. To prove this we note first that for any space Y, a continuous function $f\colon X \to Y$ induces a continuous function $F\colon I^{C(X,I)} \to I^{C(Y,I)}$ as follows. If $k \in C(Y,I)$ then $k \circ f \in C(X,I)$ so we may define $F\colon I^{C(X,I)} \to I^{C(Y,I)}$ by $F(\langle t_\lambda \rangle) = \langle (t_{k\circ f})_k \rangle$; that is, the $k \circ f$th coordinate of $\langle t_\lambda \rangle$ is taken as the kth coordinate of $F(\langle t_\lambda \rangle)$. Since $p_k \circ F = p_{k\circ f}$ is a continuous map for each k, F itself is a continuous map. Now we show that $F \circ h_X = h_Y \circ f$ by computing the kth coordinate: $(F \circ h_X(x))_k = (h_X(x))_{k\circ f} = k \circ f(x) = (h_Y(f(x)))_k$. Thus $F(h_X(X)) \subset h_Y(Y)$ so by the continuity of F, $F(\beta X) \subset \overline{F(h_X(X))} \subset \beta Y$. Thus F restricts to $\beta f\colon \beta X \to \beta Y$ such that $\beta f \circ h_X = h_Y \circ f$. Now since Y is compact $h_Y\colon Y \to \beta Y$ is a homeomorphism so $\hat{f} = h_Y^{-1} \circ \beta f$ is the desired continuous extension of f, for $\hat{f} \circ h_X = h_Y^{-1} \circ \beta f \circ h_X = h_Y^{-1} \circ h_Y \circ f = f$. Since Y is Hausdorff, $\hat{f}$ is unique, for a map from any space X to a Hausdorff space Y is determined by its values on any dense subset of X.

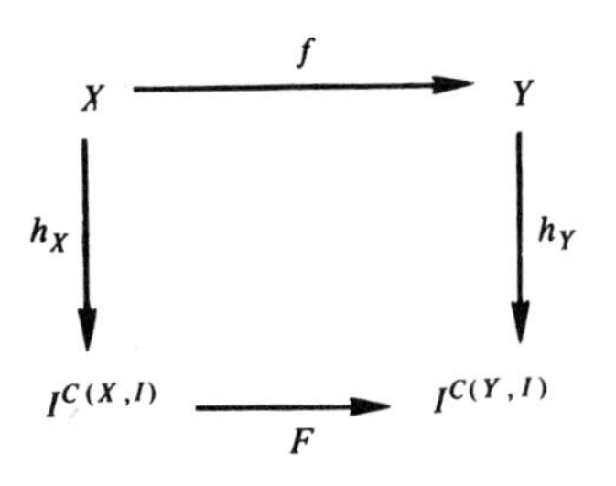

3. The properties that βX is compact, $h(X)$ is dense in βX, and

that every map from X to a compact Hausdorff space Y may be extended uniquely to βX characterize βX. For let T have those properties. Then there exists an inclusion f of X into T with a unique extension $\hat{f}: \beta X \to T$, and a unique extension $\hat{h}$ to T of the inclusion h of X into βX. Since both $\hat{h} \circ \hat{f}$ and the identity $i: \beta X \to \beta X$ are extensions to βX of $h: X \to \beta X$, $\hat{h} \circ \hat{f}$ must be the identity on βX; likewise $\hat{f} \circ \hat{h}$ is the identity on T. Hence βX and T are homeomorphic.

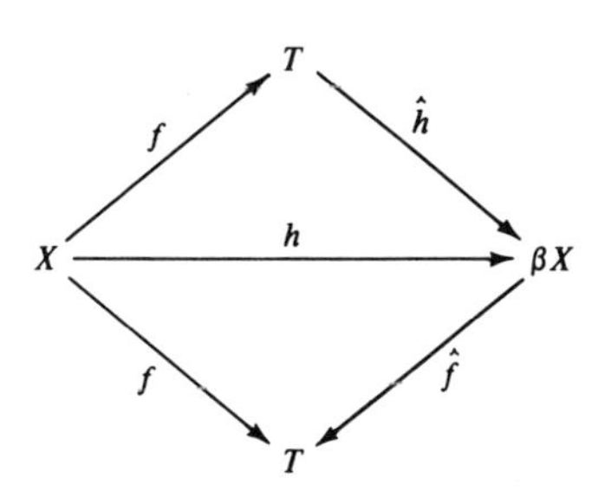

4. A second description of βX may be given as follows. We call $A \subset X$ a zero set if for some continuous real-valued function f on X, $A = \{x \in X | f(x) = 0\}$. Clearly, for a completely regular space X, the collection Z of zero-sets forms a basis for the set of closed sets of X, that is, every closed set is an intersection of zero-sets of X. We note that the points of X are in a natural correspondence with the principal ultrafilters of zero-sets and we shall identify a point with the corresponding ultrafilter; further, we shall restrict our attention to ultrafilters of zero-sets which are the intersections of ordinary ultrafilters with Z, the collection of zero-sets. We now take as a new definition of βX the set of all ultrafilters of zero-sets of X.

   With these conventions we define a topology on βX by taking as a basis of closed sets all sets of the form $C_A = \{F \in \beta X | A \in F\}$, where A is any zero-set; clearly $C_A \cap C_B = C_{A \cap B}$. Then the function $h: X \to \beta X$ which assigns to x the principal ultrafilter of all zero-sets containing x is a homeomorphism into βX and its image is dense in βX with the given topology. To see this we note that h is one-to-one, since distinct points give rise to distinct principal ultrafilters, and if A is a zero-set in X that $h(A)$ is just $h(X) \cap C_A$; thus h of each closed set is closed and h of each open set is open. Since X is the only set contained in all principal ultrafilters, $C_X = \beta X$ is the only closed set containing $h(X)$. Thus $\overline{h(X)} = \beta X$.

5. βX is compact since it satisfies the finite intersection axiom. Suppose $\{C_\lambda\}_{\lambda \in \Delta}$ is a collection of closed subsets of βX with non-empty finite intersections. Then if each C_λ is the intersection of closed basis sets $C_{A(\lambda,\alpha)}$ (that is, $C_\lambda = \bigcap_\alpha C_{A(\lambda,\alpha)}$) the family

$\{C_{A(\lambda,\alpha)}\}$ also has nonempty finite intersections. Hence so does the family $\{A(\lambda,\alpha)\}$, which means that some ultrafilter F in βX contains all of the zero-sets $A(\lambda,\alpha)$. Clearly for all α and λ, $F \in C_{A(\lambda,\alpha)} = \{G \in \beta X | A(\lambda,\alpha) \in G\}$, so $F \in \bigcap_\lambda \bigcap_\alpha C_{A(\lambda,\alpha)} = \bigcap_\lambda C_\lambda$.

6. The two spaces βX are homeomorphic since they both satisfy the three characterizing properties enumerated above. To complete the characterization, we must only show that every continuous function from X to a compact Hausdorff space Y has a unique continuous extension to (the new) βX. Since X is dense in βX any extension we can produce will be unique; so suppose $f: X \to Y$, and let $F \in \beta X$. Let $E = E(f,F) = \{$zero-sets $A \subset Y | f^{-1}(A) \in F\}$; then E is a filter in Y which inherits from F the property that whenever A and B are zero-sets for which $A \cup B \in E$, either A or B also belongs to E. Since Y is compact, $\bigcap_{A \in E} A$ is nonempty; suppose p and q are both in each set of E, $p \neq q$. Then since Y is completely regular, being compact and Hausdorff, there exist disjoint open neighborhoods U and V of p and q respectively whose complements are zero-sets. Thus since $(Y - U) \cup (Y - V) = Y \in E$ either $Y - U$ or $Y - V$ must also be in E; but neither can be in E if both p and q are in every set in E. This contradiction shows that the sets of E can have at most one, and hence precisely one point in common. This point we call $\hat{f}(F)$, and thereby define $\hat{f}: \beta X \to Y$; $\hat{f}$ is the desired extension of f. To show that $\hat{f}$ is continuous we need only show that the inverse image of any closed zero-set is closed, and this follows from the fact that for every zero-set $A \subset Y$, $\hat{f}^{-1}(A) = C_{f^{-1}(A)}$.

111. Stone-Čech Compactification of the Integers

Let (X,τ) be the Stone-Čech compactification of Z^+, the space of positive integers with the discrete topology.

1. The set X is that subset of $I^{C(Z^+,I)}$ which is the closure of the image of $h_{Z^+}: Z^+ \to I^{C(Z^+,I)}$ where h_{Z^+} is defined by $h_{Z^+}(n) = \langle \lambda(n)_\lambda \rangle$ for $\lambda \in C(Z^+,I)$. Since Z^+ is discrete, $C(Z^+,I)$, the set of all continuous functions from Z^+ to I, is merely the set of all sequences of numbers in $I = [0,1]$. Since the set of all zero-sets of functions from Z^+ to R is just the power set $P(Z^+)$ we may also describe X as the set of all ultrafilters on Z^+. Since the prin-

cipal ultrafilters correspond exactly to the points of Z^+ we may consider X to be $Z^+ \cup M$ where M is the collection of all nonprincipal ultrafilters of Z^+. In this case we have as a basis for our topology the collection of all sets of the form $U_A = \{F \in X | A \in F\}$ for $A \subset Z^+$, since $X - U_A = \{F \in X | (X - A) \in F\}$ is by definition a basis closed set. If $x \in A$, and if F_x is the principal ultrafilter containing $\{x\}$, $A \in F_x$; so, happily, $F_x \subset U_A$, which can be interpreted to mean that $A \subset U_A$. Clearly each U_A is closed as well as open; furthermore, $U_A \cap U_B = U_{A \cap B}$, and $A \subset B$ implies $U_A \subset U_B$.

2. The cardinality of $C(Z^+,I)$ is the cardinality of all countable subsets of I which is c, the cardinality of the reals. Thus card $(I^{C(Z^+,I)}) = c^c$ so card $(X) \leq c^c = 2^c$. Analogously, the set of ultrafilters is a subset of $P(P(Z^+))$ and so has cardinality less than or equal to $2^{2^{\aleph_0}} = 2^c$.

3. Since the Cartesian product of $c = 2^{\aleph_0}$ separable spaces is separable, $I^{C(Z^+,I)}$ must have a countable dense subset D. So there is a surjection $\phi: Z^+ \to D$ which is continuous since Z^+ is discrete. Thus ϕ can be extended continuously to $\hat{\phi}: X \to \bar{D}$. Since X is compact, so is $\hat{\phi}(X)$; so $\hat{\phi}$ (X) is a closed set containing D, which can only be $\bar{D} = I^{C(Z^+,I)}$. Thus card $(X) \geq$ card $(I^{C(Z^+,I)}) = 2^c$. So therefore X has cardinality 2^c.

4. When βX is considered as a set of ultrafilters the above cardinality argument takes the following form. We must construct, so to speak, $2^{2^{\aleph_0}}$ ultrafilters on the positive integers. We in fact construct 2^{2^α} ultrafilters on any infinite set with cardinal α. Let F be the set of all finite subsets f of X and let Φ denote the set of all finite subsets ϕ of F. Note that $F \times \Phi$ has the same cardinality as X so it is sufficient to construct the desired ultrafilters on $F \times \Phi$. For any subset $A \subset X$ we define $b_A = \{(f,\phi) \in F \times \Phi | A \cap f \in \phi\}$; let b_A' denote $(F \times \Phi) - b_A$. Now for each S in $P(P(X))$ let $B_S = \{b_A | A \in S\} \cup \{b_A' | A \notin S\} \subset P(F \times \Phi)$. Then B_S has the finite intersection property, and thus is the base for a filter on $F \times \Phi$. If S and T are different subsets of $P(X)$, there is some set $A \in S - T$; thus $b_A \in B_S$, and $b_A' \in B_T$, so any ultrafilters containing B_S and B_T are distinct. Thus there are at least as many ultrafilters as subsets of $P(X)$—that is, 2^{2^α} many. We note that of the 2^{2^α} ultrafilters, at most α are principal, so we have 2^{2^α} nonprincipal ultrafilters.

5. If $F \in M$, the sets U_A for all $A \in F$ form a local neighborhood basis at F. So if O is an open set and if $A = O \cap Z^+$, O must be contained in U_A, for if $F \in O \cap M$, there exists a set $B \subset Z^+$ such that $F \in U_B \subset O$. But then $B \subset A$, so $F \in U_A$.

6. If $A = O \cap Z^+$, where O is open in X, and if $F \in U_A$, every neighborhood U_B of F must intersect O since $B \cap A \neq \emptyset$. Thus $F \subset \bar{O}$. Conversely, if $G \in U_A$, then U_{Z^+-A} is an open set containing G which is disjoint from O. Thus $\bar{O} = U_A$, so X is extremally disconnected and since if $O = U_A$ we have $\bar{U}_A = U_A$, X is zero dimensional. In fact if $Z^+ = A \cup B$, $A \cap B = \emptyset$, then $U_A \cup U_B = X$ and $U_A \cap U_B = \emptyset$.

7. Every neighborhood U_A of every point $F \in M$ contains infinitely many elements of M, so M is dense-in-itself. Thus X is not scattered.

8. X is separable since Z^+ is dense in X. But, since X is extremally disconnected yet not discrete, it is not metrizable; since it is regular, it cannot be second countable.

112. Novak Space

If Z^+ denotes the positive integers with the discrete topology, and if S is the Stone-Čech compactification of Z^+, we will construct by transfinite induction a certain subset P of S. Let F be the family of all countably infinite subsets of S, well ordered by the least ordinal Γ of cardinal $2^c = \text{card}\,(S)$. Let $\{P_A | A \in F\}$ be a collection of subsets of S such that card $(P_A) < 2^c$, $P_D \subset P_A$ whenever $D < A$, and $\hat{f}(P_A) \cap P_A = \emptyset$ where $\hat{f}$ is the unique extension to $S = \beta(Z^+)$ of the continuous function $f\colon Z^+ \to Z^+$ which permutes each odd integer with its even successor: $f(n) = n + (-1)^{n+1}$. Then we define $P = \bigcup\{P_A | A \in F\}$, and then define Novak's space by $X = P \cup Z^+$; X is a subspace of $S = \beta(Z^+)$.

1. The collection $\{P_A | A \in F\}$ can be defined inductively as follows: if $B \in F$ where P_A has been defined for all $A < B$, let $Q_B = \bigcup\{P_A | A < B\}$. Then card $(Q_B) < 2^c$ since card $\{A | A < B\} < 2^c$ and card $(P_A) < 2^c$. Furthermore, since $f \circ f$ is the identity on Z^+, so is $\hat{f} \circ \hat{f}$ on S. Thus $\hat{f}$ is invertible, so $\hat{f}(Q_B) = \bigcup \hat{f}(P_A)$ cannot intersect Q_B. Now $\bar{B}$ is an infinite closed subset of S, and any such set must have cardinality 2^c; hence card $\bar{B} >$ card $B \cup \hat{f}(Q_B)$, so there exists a point $x \in \bar{B} - B$ such that $x \notin \hat{f}(Q_B)$. Let $P_B = Q_B \cup \{x\}$. Then clearly card $(P_B) < 2^c$;

furthermore, $\hat{f}(P_B) \cap P_B = \hat{f}(Q_B \cup \{x\}) \cap (Q_B \cup \{x\}) = \{\hat{f}(x)\} \cap Q_B$ since $\hat{f}$ leaves no point fixed. But this intersection is empty since otherwise $x = \hat{f}(\hat{f}(x)) \in \hat{f}(Q_B)$. This completes the inductive construction of the sets P_A.

2. If B is a countably infinite subset of S, P contains a limit point of B, since by construction P contains a point of $\bar{B} - B$ for each such B. So X is countably compact.

3. Let $K = \{(n,\hat{f}(n)) | n \in Z^+\}$. Since $\hat{f}$: $S \to S$ is continuous, its graph $G = \{(x,\hat{f}(x)) \in S \times S\}$ is closed in $S \times S$. Since $P \cap \hat{f}(P) = \varnothing$ by construction and $\hat{f}(Z^+) \subset Z^+$, $G \cap (X \times X) = K$, so K is closed in $X \times X$. Furthermore, K is infinite and contains no limit points of itself since it is the graph of the homeomorphism $\hat{f}$ on the discrete set Z^+. So K is an infinite set of $X \times X$ without a limit point so $X \times X$ is not countably compact.

4. X is separable since $\overline{Z^+} = X$, and X is completely regular since it is a subspace of the normal space S.

113. Strong Ultrafilter Topology

Let Z^+ be the positive integers, and let M be the collection of all non-principal ultrafilters on Z^+. Let $X = Z^+ \cup M$, and let the topology τ on X be generated by the points of Z^+ together with all sets of the form $A \cup \{F\}$ where $A \in F \in M$.

1. X is Hausdorff, since any two members F and G of M, being ultrafilters, are incomparable. So there exist $A \in F - G$, $B \in G - F$. Then since F is an ultrafilter, $B' \in F$, so $A \cap B' = A - B \in F$. Similarly, $B - A \in G$, and so $(A - B) \cup \{F\}$ and $(B - A) \cup \{G\}$ are disjoint neighborhoods of F and G. (Note that $F \in M$ can be separated from any $y \in Z^+$ precisely since no y can be contained in all sets of F since F can have no cluster points.)

2. We can prove that X is extremally disconnected by showing that if O is an open subset of X, $\bar{O}$ is open. Suppose p is a limit point of O which does not belong to O; since each point of Z^+ is open, $p \in X - Z^+ = M$. So p is an ultrafilter, say F, and every neighborhood $A \cup \{F\}$ of p (where $A \in F$) intersects O. But since F itself does not belong to O, this intersection is contained in Z^+. Thus, $O \cap Z^+$ intersects every member of the ultrafilter F; but it is a property of ultrafilters that for every subset S

(of Z^+), either S or its complement belongs to the ultrafilter. Since $O \cap Z^+$ does not intersect its own complement, $O \cap Z^+$ itself must belong to the ultrafilter F. That is, $(O \cap Z^+) \in p$! Thus $(O \cap Z^+) \cup \{F\}$, or equivalently, $(O \cap Z^+) \cup \{p\}$, is open. Thus $O \cup \{p\} = O \cup ((O \cap Z^+) \cup \{p\})$ is open, and since p was an arbitrary limit point of O, $\bar{O}$ must be open. Thus X is extremally disconnected.

3. Any basis element $A \cup \{F\}$ has as a limit point every ultrafilter G which contains A as an element, for if $B \in G$ and $A \in G$, then $A \cap B \neq \emptyset$, so $B \cup \{G\} \cap A \cup \{F\} \neq \emptyset$. So if $B \subset A$, $\overline{B \cup \{F\}}$ contains all ultrafilters which contain B, which means that $\overline{B \cup \{F\}}$ is not contained in $A \cup \{F\}$. Thus X cannot be T_3, so it is not zero dimensional.

4. X is scattered since it cannot contain any nonempty dense-in-itself subsets. For no dense-in-itself set can contain a point of Z^+, yet no point of M can be a limit point of a subset of M. In fact, Z^+ is discrete, and in the induced topology, so is M. Clearly M is an infinite subset of X with no limit point, so X is not countably compact.

5. Since every open set in X contains an integer, Z^+ is dense in X. Thus X is separable. But M is uncountable, so the collection of all open sets $A \cup \{F\}$ where $F \in M$ is an uncountable open covering of X which has no countable subcover. Thus X is not Lindelöf.

6. Since X is extremally disconnected it is totally separated, and thus Urysohn.

7. The direct sum of X with $\beta(Z^+)$, the Stone-Čech compactification of the integers, is extremally disconnected, but neither zero dimensional nor scattered, since both spaces are extremally disconnected, but one is not zero dimensional and the other is not scattered.

114. Single Ultrafilter Topology

Let $X = Z^+ \cup \{F\}$ where F is a nonprincipal ultrafilter on Z^+. We take as a basis of open sets all sets of the form $A \cup \{F\}$ where $A \in F$, together with the points of Z^+.

1. X is Hausdorff for clearly any two points of Z^+ may be separated, while if $x \in Z^+$, then since F is nonprincipal, $\{x\} \notin F$; thus $(Z^+ - \{x\}) \cup \{F\}$ is a neighborhood of F disjoint from the open set $\{x\}$.

2. X is extremally disconnected for the only limit point of any set is F, but F is a limit point of A iff $A \in F$ and then $A \cup \{F\}$ is open. Thus, also, X is zero dimensional since the sets $A \cup \{F\}$ are both open and closed.

3. X is clearly scattered, since no subset can be dense-in-itself, but not discrete since the point F is not open.

115. Nested Rectangles

In the Euclidean plane, let L_1 designate the line $x = 1$, L_2 the line $x = -1$, and R_n the boundary of rectangles centered at the origin, of height $2n$ and width $2n/(n + 1)$. Let $X = L_1 \cup L_2 \cup (\cup R_n)$, and let X inherit the topology from the Euclidean plane.

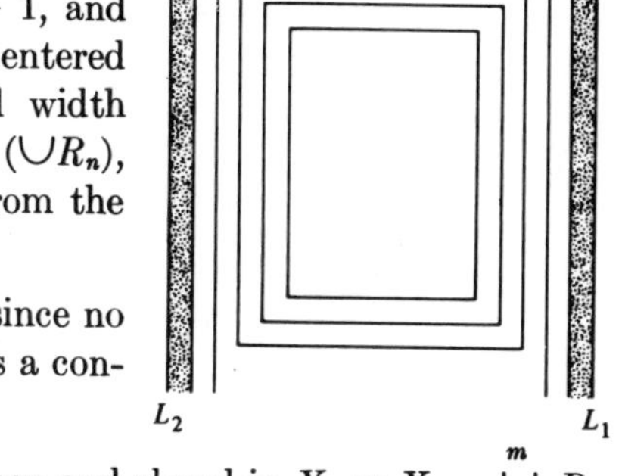

1. X is not locally connected since no point on either L_1 or L_2 has a connected neighborhood.

2. Each rectangle R_n is both open and closed in X, so $X - \bigcup_{n=1}^{m} R_n$ is also open and closed. Thus $L_1 \cup L_2 = X - \cup R_n$ is a quasi-component of X. But L_1 is a component of X, since it is connected and no larger subset of X is connected.

116. Topologist's Sine Curve

117. Closed Topologist's Sine Curve

118. Extended Topologist's Sine Curve

Let S be the graph of $f(x) = \sin(1/x)$ for $0 < x \leq 1$, considered as a subset of the Euclidean plane with the induced topology. The topologist's sine curve is the set $S \cup \{(0,0)\}$ which we will denote by S^*.

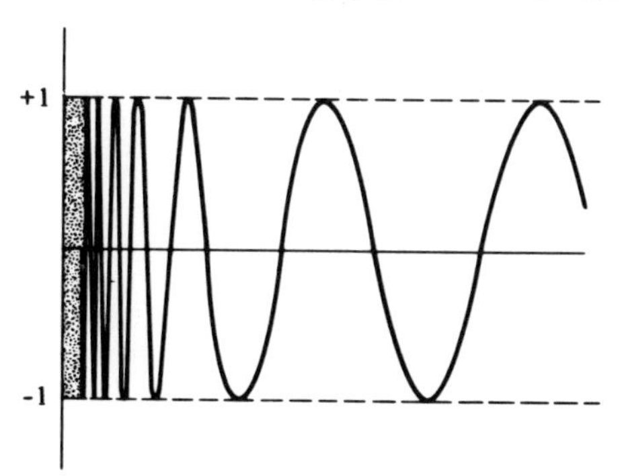

1. S^* is not locally compact, since the point (0,0) has no compact neighborhood. For any neighborhood N of (0,0) contains a set $\Delta \cap S^*$, where Δ is a disc centered at the origin of radius ϵ. Then any horizontal line passing through, say $(0,\epsilon/2)$ intersects $\Delta \cap S^*$ in a sequence of points which has no accumulation point in N. So N cannot be countably compact, and thus not compact.

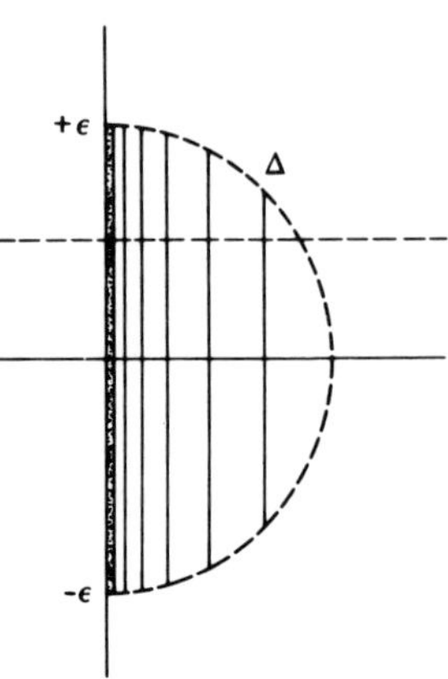

2. The map $f: \{-1\} \cup (0,1] \to S^*$ defined by $f(-1) = (0,0)$, $f(x) = (x, \sin(1/x))$ for $x \in (0,1]$ is continuous, so the continuous image of a locally compact space need not be locally compact.

3. The closed topologist's sine curve $\bar{S}$, which is $S \cup \{(0,y)|-1 \leq y \leq 1\}$, is compact, being closed and bounded, as well as connected, for $\bar{S}$ is the closure of the continuous image of the connected set (0,1]. Since $S = f((0,1])$ is connected, and $S \subset S^* \subset \bar{S}$, S^* is also connected.

4. Clearly neither S^* nor $\bar{S}$ is locally connected. But any continuous function from the locally connected compact set [0,1] to the Hausdorff space S^* (or $\bar{S}$) must have a locally connected and connected image. Thus no path can join the point (0,0) to $(4\pi,0)$ in either $\bar{S}$ or S^*, so neither space is path connected.

5. $\bar{S}$ has two path components, S and $L = \{(0,y)|-1 \leq y \leq 1\}$. Though L is closed, S is not; but $\bar{S}$ is not path connected, though S is. Similarly, S^* has two path components, S and $\{(0,0)\}$.

6. The extended topologist's sine curve $T = \bar{S} \cup \{(x,1)|0 \leq x \leq 1\}$ is arc connected, but not even locally connected.

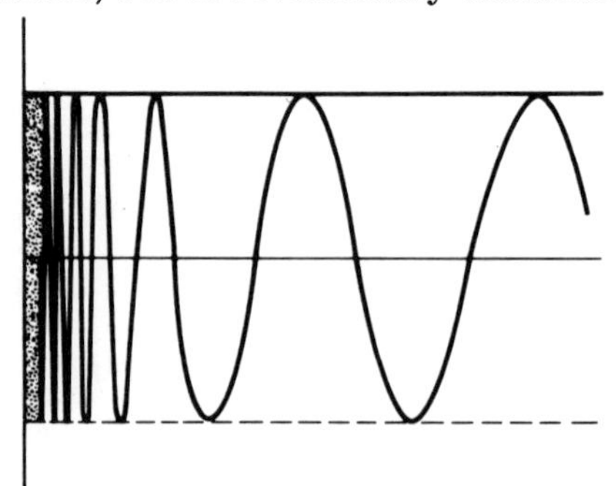

119. The Infinite Broom

120. The Closed Infinite Broom

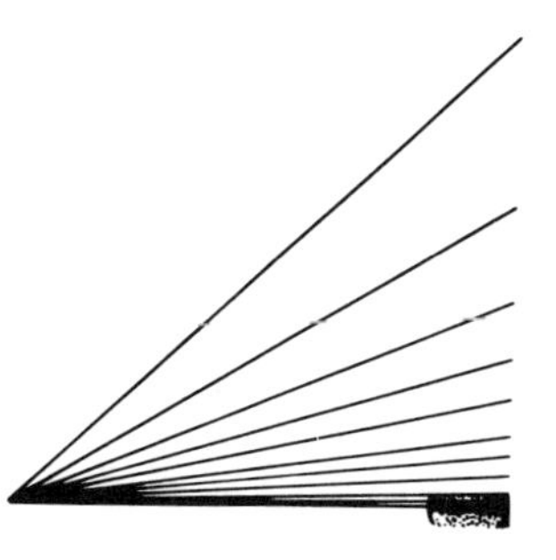

In the Euclidean plane, the infinite broom B is the union of the closed line segments joining the origin to the points $\{(1,1/n)|n = 1, 2, 3, \ldots\}$ together with the half-open interval $(\frac{1}{2},1]$ on the x axis. The closed infinite broom is then $\bar{B}$, the union of B and the interval $(0,1]$.

1. B is connected since the line segments through the origin all have a common point, and every open set in the plane which contains $(\frac{1}{2},1]$ intersects these line segments. Thus $\bar{B}$ is also connected.
2. Neither B nor $\bar{B}$ is locally connected since every small open neighborhood of the point $(\frac{3}{4},0)$ has a separation.
3. Clearly $\bar{B}$ is arc connected, yet B is not even path connected, for any path connecting a point of $(\frac{1}{2},1]$ to a point off the x axis would be a continuous map from a locally connected compact space (namely $[0,1]$) onto a Hausdorff space which was not locally connected (namely $f[0,1]$).
4. A more interesting infinite broom may be formed by joining a sequence of closed brooms end to end as pictured. Since no open set containing the point $(0,0)$ is connected (except for X itself),

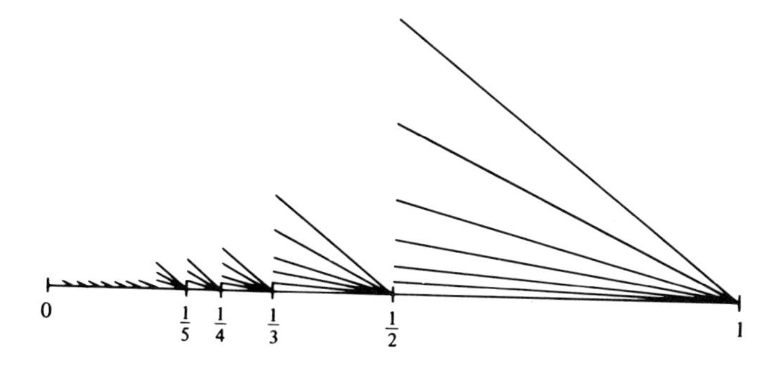

this space fails to be locally connected at the point $(0,0)$ even though this point has a basis of open sets whose closures are connected.

121. The Integer Broom

Let X be the set of points with polar coordinates $\{(n,\theta)\}$ in the plane R^2 where n is a nonnegative integer and $\theta \in \{1/n\}_1^\infty \cup \{0\}$. We define a topology τ on X by taking as a basis of open sets all sets of the form $U \times V$ where U is an open set in the right order topology on the nonnegative integers and V is open in $\{0\} \cup \{1/n\}_1^\infty$ in the topology induced from the reals. The only neighborhood of the origin is X itself.

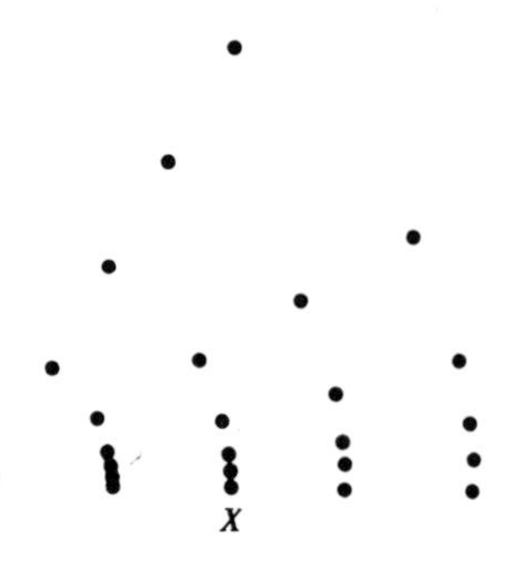

1. X is clearly T_0, but neither T_1 nor T_2; (X,τ) is compact since the only open set containing the origin is X itself.

2. X is not locally connected since $(1,0)$ does not have a basis of connected neighborhoods.

3. Since X is countable it is not arc connected yet it is path connected for the function f: $[0,1] \to X$ which maps the interval $[0,\frac{1}{2})$ to the point (n_1,θ_1), $(\frac{1}{2},1]$ to (n_2,θ_2) and the point $\frac{1}{2}$ to the origin is a path joining (n_1,θ_1) and (n_2,θ_2).

122. Nested Angles

Let X be the subset of the plane E^2 consisting of line segments joining the points $(0,1)$ and $(n,1/(n+1))$ for $n \in Z^+$; the half-lines $y =$

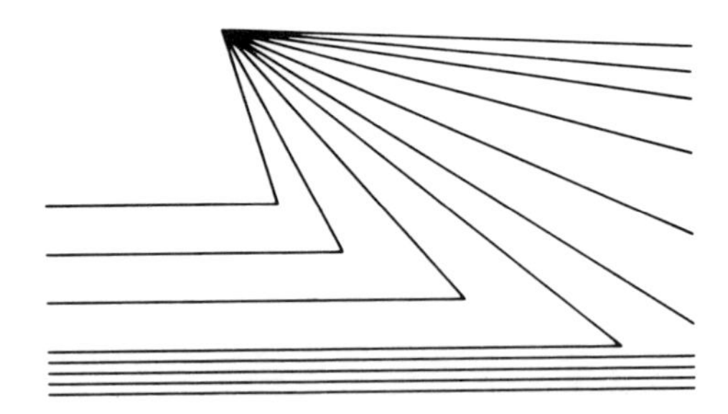

$1/(n+1)$, $n \in Z^+$, $x \leq n$; and the line $y = 0$. We give X the induced topology.

1. X is the closure of a family of (bent) lines with the point $(0,1)$ in common. Hence it is connected.

2. Since X is a closed subset of E^2 it is locally compact, but it is not compact since it is not bounded.

3. The set $X - \{(0,1)\}$ is not connected; in fact, each angle and the x axis are components. So, in particular, X is not locally connected.

123. The Infinite Cage

The infinite cage X is the union of three types of sets:

$$A_n = \{(1/n,y,0) \in R^3 | 0 \leq y \leq 3n\},$$
$$B_n = \{(0,y,0) \in R^3 | 2n - \tfrac{1}{2} \leq y \leq 2n + \tfrac{1}{2}\},$$
$$C_n = \{(x,y,z) \in R^3 | 0 \leq x \leq 1/n,\ y = 2n,\ z = x(1/n - x)\}.$$

We define X to be $\bigcup_{n=1}^{\infty} (A_n \cup B_n \cup C_n)$ and give it the induced Euclidean topology.

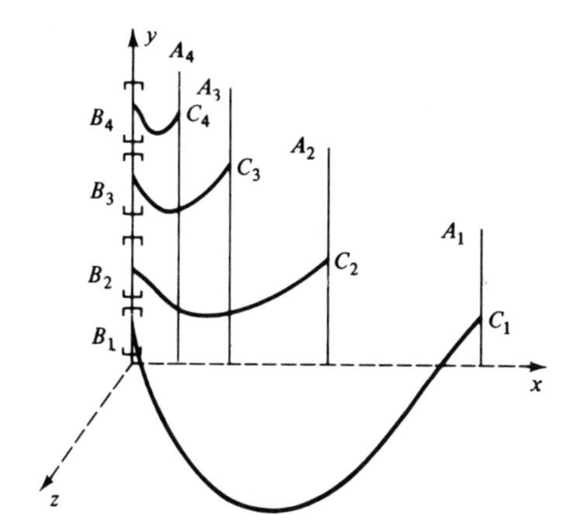

1. If $D_n = A_n \cup B_n \cup C_n$, $\{D_n\}$ is a collection of pairwise disjoint compact connected subsets of X.

2. The cage $X = \cup D_n$ is itself connected, for suppose Y,Z were a separation of X. Then each of the sets Y and Z must contain entirely each D_n which they intersect. So at least one of the sets Y or Z contains infinitely many of the sets A_i; suppose it is Y. Then Y must also contain $\cup B_n$ since any point of B_n is a limit point of any infinite collection of A_i, hence each $D_n \subset Y$. But this means that $Y = X$.

124. Bernstein's Connected Sets

Let $\{C_\alpha | \alpha \in [0,\Gamma)\}$ be the collection of all nondegenerate closed connected subsets of the Euclidean plane R^2 well ordered by Γ, the least ordinal equivalent to c, the cardinal of the continuum. We define by transfinite induction two nested sequences $\{A_\alpha\}_{\alpha<\Gamma}$ and $\{B_\alpha\}_{\alpha<\Gamma}$ such that $A_\alpha \cap B_\beta = \emptyset$ for all pairs α, β. A_1 and B_1 are merely distinct singletons selected from C_1; if $\{A_\alpha\}_{\alpha<\beta}$ and $\{B_\alpha\}_{\alpha<\beta}$ have been defined, the cardinal of $\bigcup_{\alpha<\beta} (A_\alpha \cup B_\alpha)$ is less than c, but card $C_\beta = c$. Thus we can select points $a_\beta, b_\beta \in C_\beta - \bigcup_{\alpha<\beta} (A_\alpha \cup B_\alpha)$, and define $A_\beta = \{a_\beta\} \cup (\bigcup_{\alpha<\beta} A_\alpha)$, $B_\beta = \{b_\beta\} \cup (\bigcup_{\alpha<\beta} B_\beta)$. Let $A = \bigcup_{\alpha<\Gamma} A_\alpha$, and $B = R^2 - A$.

1. A and B are clearly disjoint subsets of R^2 containing $\bigcup A_\alpha$ and $\bigcup B_\alpha$ respectively and any nondegenerate closed connected subset of R^2 must intersect both A and B.

2. Every open subset of R^2 contains some nondegenerate closed connected set, so must intersect both A and B. Thus both A and B are dense in R^2.

3. Suppose A were separated by the disjoint open sets U and V; $A \cap U \neq \emptyset$, $A \cap V \neq \emptyset$, $A \subset (U \cup V)$. The complement in R^2 of $U \cup V$ separates the plane, so must contain a nondegenerate closed connected set C. But then A would intersect C even though $C \subset X - \mathrm{A}$. This contradiction shows that A (and, similarly, B) is connected.

125. Gustin's Sequence Space

Gustin's sequence space (X,τ) is constructed from the set $X = Y \cup (Z^+ \times W)$ where Y consists of all finite sequences of positive integers having an even number of terms (including the null sequence denoted by 0), Z^+ is, as usual, the positive integers, and W is the collection of all unordered pairs (that is, all subsets of size two) from Y. Now if α and β are arbitrary finite sequences, we will denote by $\alpha\beta$ the sequence formed by adjoining β to the end of α, by $\alpha \geq i (i \in Z^+)$ the condition that $a \geq i$ for all $a \in \alpha$, and by $\beta \supset_i \alpha$ the existence of a sequence $\gamma \geq i$ such that $\beta = \alpha\gamma$. For any sequence α, let $U_i(\alpha) = \{\beta \in Y | \beta \supset_i \alpha\}$.

Before defining the topology on X, we select some one-to-one correspondence p between the countable set W and the set of positive prime numbers. Then we define $q\colon (Z^+ \times W) \to Z^+$ by $q(n,w) = [p(w)]^n$. Finally, we define the topology τ on X by selecting the set $U_i(\alpha)$ as open neighborhoods of the point $\alpha \in Y$, and $V_i(n,w) = \{(n,w)\} \cup$

$U_i(\alpha q(n,w)) \cup U_i(\beta q(n,w))$ as open neighborhoods of the point $(n,w) = (n,\{\alpha,\beta\}) \in Z^+ \times W$.

1. (X,τ) can be shown Hausdorff by a straightforward consideration of cases. If $\alpha,\beta \in Y$, $\alpha \neq \beta$, then we can find an integer n larger than any term of α or β. Clearly $U_n(\alpha) \cap U_n(\beta) = \emptyset$. Suppose $\gamma \in Y$ and $(n,w) = (n,\{\alpha,\beta\}) \in Z^+ \times W$; select an integer m greater than every term of the sequences γ, $\alpha q(n,w)$, $\beta q(n,w)$; then $U_m(\gamma) \cap V_m(n,w) = \emptyset$, for γ, having an even number of terms can never equal $\alpha q(n,w)$ or $\beta q(n,w)$ since they each have an odd number of terms. Finally, if $(n,w) = (n,\{\alpha,\beta\})$ and $(m,z) = (m,\{\gamma,\delta\})$ are distinct points in $(Z^+ \times W)$, we select an integer i greater than any term in $\alpha q(n,w)$, $\beta q(n,w)$, $\gamma q(m,z)$ or $\delta q(m,z)$. Then $V_i(n,w)$ could intersect $V_i(m,z)$ only if one of the points $\alpha q(n,w)$ or $\beta q(n,w)$ equaled one of the points $\gamma q(m,z)$, $\delta q(m,z)$. But this could happen only if $q(n,w) = q(m,z)$, and this would mean $(n,w) = (m,z)$.
2. If $U_i(\gamma)$ is a neighborhood of $\gamma \in Y$, let $Z(i,\gamma) = \{(n,\{\alpha,\beta\}) \in Z^+ \times W | \alpha q(n,w) \supset_i \gamma$ or $\beta q(n,w) \supset_i \gamma\}$. Then clearly every point of $Z(i,\gamma)$ is a limit point of $U_i(\gamma)$; in fact, $\overline{U_i(\gamma)} = U_i(\gamma) \cup Z(i,\gamma)$.
3. If $\gamma,\delta \in Y$, $Z(i,\gamma) \cap Z(j,\delta) \neq \emptyset$ for all $i,j \in Z^+$; this is so because we can always find a point $(n,w) = (n,\{\alpha,\beta\})$ such that $q(n,w) > \max(i,j)$, and $\alpha \supset_i \gamma$ and $\beta \supset_i \delta$. Thus every two disjoint open sets in (X,τ) have closures with nonempty intersection. This shows that (X,τ) is connected.
4. Let $X^* = Y \cup \{(n,\{\alpha,0\}) | n \in Z^+, \alpha \neq 0\}$. Then every relatively open neighborhood of a point in X^* is open in (X,τ), so an argument similar to that given above shows that X^* is also connected. But the point $0 \in Y$ is a dispersion point for X^*, since $X^* - 0$ is totally disconnected.

126. Roy's Lattice Space

127. Roy's Lattice Subspace

Let $\{C_i\}_{i=1}^{\infty}$ be a countable collection of disjoint dense subsets of the rationals Q; we construct the space X by joining to $\{(r,i) \in Q \times Z^+ | r \in C_i\}$ an ideal point ω. Neighborhoods of the points of the form $(r,2n)$—that is, of points on the even numbered lines—are ordinary open intervals $U_\epsilon(r,2n) = \{(t,2n) \,\big|\, |t - r| < \epsilon\}$. But a neighborhood of

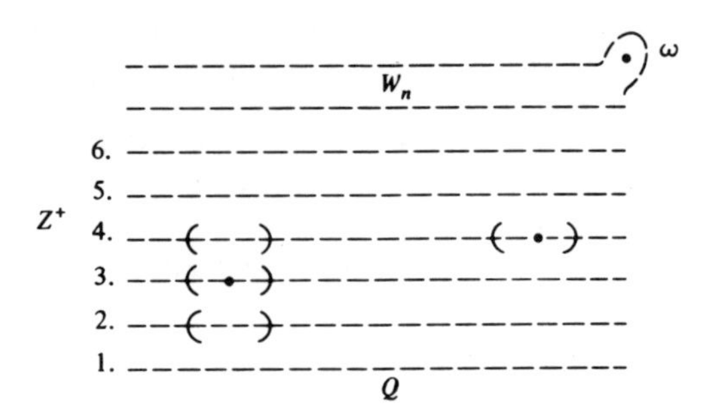

a point of the form $(r,2n-1)$ is a stack of three open intervals $V_\epsilon(r,2n-1) = \{(t,m) \mid |t-r| < \epsilon, \quad m = 2n-2, \quad 2n-1, \quad 2n\}$. A basis neighborhood of the point ω consists of all lines numbered $\geq 2n$: $W_n(\omega) = \{(s,i) \in X | i \geq 2n\}$. These neighborhoods form a basis for a topology τ on the countable set X. The subspace $X - \{\omega\}$ will be denoted by X^*.

1. Any closed set containing an even numbered line must contain both adjacent odd numbered lines since every neighborhood of every point on the line $\{(r,i) \in X | i = 2n-1\}$ intersects the lines $\{(r,i) \in X | i = 2n\}$ and $\{(r,i) \in X | i = 2n-2\}$. Similarly, every open set containing an odd numbered line must contain both adjacent even numbered lines.

2. Suppose A is an open and closed subset of X which contains ω. Then A contains a neighborhood $U = \{(r,i) \in X | i \geq 2n\}$ of ω. Since A is closed, it must contain the next lower odd numbered line; since it is open it contains the next even numbered line. Clearly it must therefore contain all lines numbered below $2n$, and thus all of X. Thus X is connected.

3. The ideal point ω is a dispersion point of X, for $X^* = X - \{\omega\}$ is totally separated and thus totally disconnected. For whenever (r,i) and $(s,j) \in X^*$, where $r < s$, we can find an irrational number t between r and s which yields a separation of X: $\{(r,i) \in X | r < t\}$ and $\{(r,i) \in X | r > t\}$.

4. Since X^* is totally separated, X^* must be Urysohn; but since X is connected and countable, X cannot be Urysohn since the image of X in $[0,1]$ would be connected and countable, which is impossible, unless the image is one point.

5. However, X is completely Hausdorff, since no two points in X have the same first coordinate. Thus we can find sufficiently

short intervals around the points with disjoint closures. Furthermore, X^* is not regular since the neighborhood $U_\epsilon(r,2n)$ cannot contain any closed neighborhood whatsoever. Thus X^* is not zero dimensional.

6. X^* contains no isolated points, so it is dense-in-itself; thus X^* is not scattered. Neither is it extremally disconnected for $\overline{U_\epsilon(r,2n)}$ is not open since it contains points of the form $(s,2n+1)$ and $(t,2n-1)$ but does not contain any of their neighborhoods.
7. If $V_\epsilon(r,2)$ is a neighborhood of the point $(r,2)$, $\overline{V_\epsilon(r,2)}$ contains points of the form $(s,1)$ which are interior points of $\overline{V_\epsilon(r,2)}$. Thus $\overline{V_\epsilon(r,2)}^\circ \neq V_\epsilon(r,2)$, so X is not semiregular.

128. Cantor's Leaky Tent

129. Cantor's Teepee

Let C be the Cantor set situated on the unit interval $[0,1]$; let p be the point $(\frac{1}{2},\frac{1}{2})$ in the coordinate plane. Let X be the cone over C with vertex at p. That is, if $L(c)$ denotes the line segment joining p to the point $c \in C$, $X = \bigcup\{L(c) | c \in C\}$. If E denotes the subset of C consisting of the endpoints of the deleted intervals, we let X_E denote the cone over E: $X_E = \bigcup\{L(c) | c \in E\}$; similarly, if $F = C - E$, X_F denotes the cone over F. Then we define $Y_E = \{(x,y) \in X_E | y \in Q\}$,

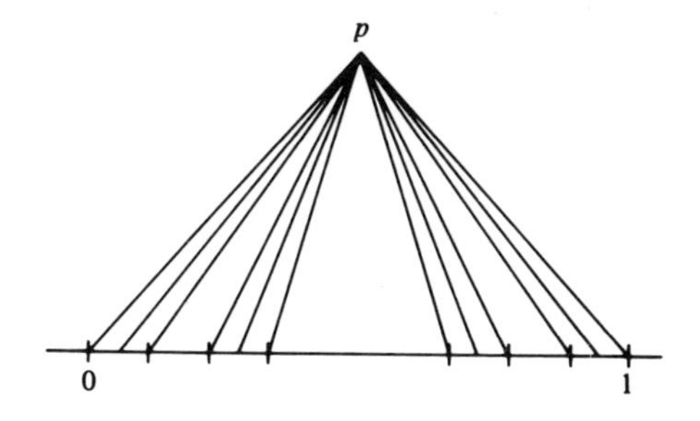

where Q denotes the rationals, $Y_F = \{(x,y) \in X_F | y \notin Q\}$, and $Y = Y_E \cup Y_F$. Both X and Y carry the induced Euclidean topology.

1. To prove Y connected we consider a separation A,B where $p \in A$. We will show that for some dense set $S \subset C$, A contains all the points of Y which lie in the cone over S (except those for which $y = 0$); thus $\bar{A} = Y$. For each $c \in C$, we let $l(c)$ be the least upper bound of $B \cap L(c)$; if $B \cap L(c) = \varnothing$, we let $l(c) = c$.

Then $l(c) \notin Y$ unless $l(c) = c$, for otherwise it would be a limit point of both A and B. Furthermore, $l(c) = c$ can be in $Y = Y_E \cup Y_F$ only if it is in Y_E, or equivalently, in E—for the y coordinate of $l(c)$, namely 0, is rational. Thus for each $c \in C$, either $l(c) \notin Y$, or $l(c) = c \in Y_E$.

Let $S = \{c \in F | c = l(c)\}$, and let $T_i = \{c \in C | L(c) \cap \overline{H}_i \neq \emptyset\}$ where, if $\{r_i\}$ is an enumeration of the rationals in $(0,\frac{1}{2}]$, $\overline{H}_i = \{(x,r_i) | l(c) = (x,r_i)$ for some $c \in F\}$. Each $\overline{H}_i$ is a closed, bounded subset of the line $y = r_i$, and T_i is the continuous projection through p of $\overline{H}_i$; so each T_i is closed. Furthermore, $\overline{H}_i \cap L(c) = \emptyset$ for every $c \in E$ and each $i > 0$, for if $(x,y) \in \overline{H}_i \cap L(c)$ where $c \in E$, y is rational and $(x,y) \in X_E$; so $(x,y) \in Y_E \subset Y = A \cup B$. But $H_i \subset \bar{A} \cap \bar{B}$, so $\overline{H}_i \cap L(c) \subset (\bar{A} \cap \bar{B}) \cap (A \cup B) = \emptyset$.

Thus each $T_i \subset F$, so if $T = \cup T_i$, $F = S \cup T$; for if $c \in F$, $l(c)$ must be rational—since otherwise $l(c) \in Y_F \subset Y$, which is impossible if $c \in F$. Thus $C = E \cup S \cup T$, where C is a complete metric space, E is countable and each T_i is nowhere dense in C since $T_i = \overline{T}_i$, and the interior of T_i in C is empty (otherwise we could find an open interval U such that $U \cap C \subset T_i \subset F$; but in each such $U \cap C$ there must be a point of E). So $E \cup T$ is of the first category, but C is not. Furthermore, no open subset of C is of the first category in C, so each such set must contain a point of $S = C - (E \cup T)$; thus S is dense in C.

Now suppose $q \in B$; then, since S is dense in C, every open set containing q intersects a segment of $L(c)$ for some $c \in S$. But by definition of S, the set $Y \cap (L(c) - \{c\})$ is contained in A whenever $c \in S$; thus $q \in \bar{A}$, so $\bar{A} = Y$. This proves that Y is connected.

2. The point $p = (\frac{1}{2},\frac{1}{2})$ is a dispersion point of Y, for each point of $Y^* = Y - \{p\}$ is a component of Y^*. For suppose A is a connected subset of Y^*, then clearly A must lie entirely within one line $L(c)$, for otherwise some line through p would separate A. But $L(c) \cap Y^*$ is totally disconnected, so A can contain at most one point.

3. But Y^* is not totally separated, for the lines $L(c)$ are the quasi-components of Y^*. To see this we observe that each line $L(c)$ is the intersection of cones over certain intervals containing c which are open and closed in Y^*. Furthermore, each open and closed set A containing a point $r \in L(c) \cap Y^*$ must contain $L(c) \cap Y^*$, since otherwise there would be a point $q \in L(c) \cap Y^*$ and a disc

$B_\epsilon(q)$ in the complement of A. Assume that q lies above r in $L(c)$; a similar argument holds if q is below r. We can then find an interval (s,t) containing c such that the cone T over (s,t) is open and closed, and separated by $B_\epsilon(q)$. Then that part of $A \cap T$ which lies below $B_\epsilon(q)$ is an open and closed subset of Y^* whose complement together with p is an open and closed subset of Y. But this is impossible since Y is connected, so A must contain all of $L(c) \cap Y^*$.

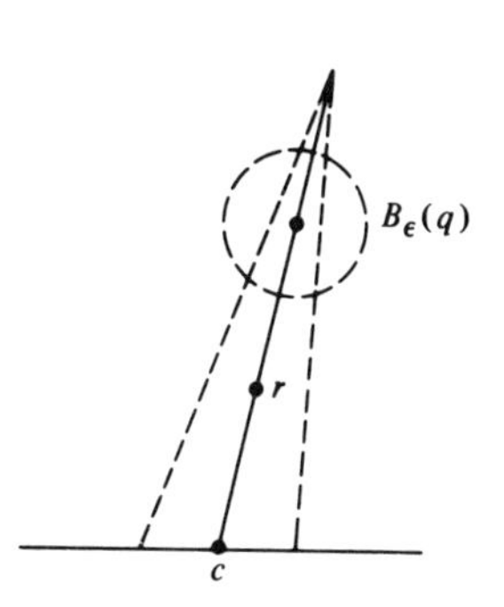

4. Clearly Y contains no nondegenerate compact connected proper subsets so is punctiform.
5. Since no point of Y^* is isolated, Y^* is not scattered.
6. The direct sum of Y^* with modified Fort space is totally disconnected but neither scattered nor Hausdorff, since modified Fort space is scattered and totally disconnected but not Hausdorff.

130. A Pseudo-Arc

By a chain $\mathfrak{D}$ in the Euclidean plane we will mean a finite collection of open sets $\{D_i\}_1^n$ (called links) such that $D_i \cap D_j = \emptyset$ iff $|i - j| > 1$. A pseudo-arc joining two points a,b in the plane is any set in R^2 resulting from the following inductive construction. Let $\{\mathfrak{D}_i\}$ be a sequence of chains such that

(i) The diameter of each open set in $\mathfrak{D}_i$ is less than $1/i$.
(ii) The closure of each link of $\mathfrak{D}_{i+1}$ is contained in some link of $\mathfrak{D}_i$.
(iii) $\mathfrak{D}_{i+1}$ is crooked in $\mathfrak{D}_i$, that is, if $D_m^{i+1}, D_n{}^{i+1} \in \mathfrak{D}_{i+1}$ with $m < n$ and $D_m^{i+1} \subset D_h^i$, $D_n^{i+1} \subset D_k^i$ with $|k - h| > 2$ then there exist D_s^{i+1}, $D_t^{i+1} \in \mathfrak{D}_{i+1}$ with $m < s < t < n$ such that D_s^{i+1} is contained in a link of $\mathfrak{D}_i$ adjacent to D_k^i and similarly D_t^{i+1} is contained in a link adjacent to D_h^i.
(iv) a is in the first link of each chain $\mathfrak{D}_i$ and b is in the final link of each chain.

If $\mathfrak{D}_i{}^* = \bigcup\limits_k D_k^i$ denotes the set of all elements of elements of $\mathfrak{D}_i$, then $X = \bigcap \overline{\mathfrak{D}_i{}^*}$ with the induced topology is a pseudo-arc.

1. Since X is an intersection of closed sets, it is closed. Since each

element of $\mathfrak{D}_1$ has diameter less than 1 and $\mathfrak{D}_1$ has only finitely many links $X \subset \mathfrak{D}_1^*$ is bounded. Thus X is compact.

2. X is connected. Suppose C_1 and C_2 are two disjoint closed (and thus compact) subsets. Then for some i, $3/i$ is less than $\inf \{d(x,y) | (x,y) \in C_1 \times C_2\}$ where d is the induced Euclidean metric on X. Thus there is some link $\mathfrak{D}_i$ whose closure intersects neither C_1 or C_2, but since the closure of every link of every chain contains a point of X, $C_1 \cup C_2 \neq X$.

3. X contains no decomposable subcontinuum, that is, X is hereditarily indecomposable. Let $Y \subset X$ be any subcontinuum and let Y be the sum of two distinct proper subcontinua H and K. Then there are points $p \in H$, $q \in K$ and an integer j such that the distance from q to H and p to K are both greater than $2/j$. Let $\mathfrak{D}_j' = \{D_h^j, D_{h+1}^j, \ldots, D_{k-1}^j, D_k^j\}$ and $\mathfrak{D}_{j+1}' = \{D_u^{j+1}, D_{u+1}^{j+1}, \ldots, D_{v-1}^{j+1}, D_v^{j+1}\}$ be subchains of $\mathfrak{D}_j$ and $\mathfrak{D}_{j+1}$ from p to q. Without loss of generality assume $p \in D_h^j$. Then D_{h+1}^j, contains no point of K, therefore it must contain a point of H; similarly, D_{k-1}^j contains no point of H but some point of K. Now since $\mathfrak{D}_{j+1}$ is crooked in $\mathfrak{D}_j$ there are links D_r^{j+1}, D_s^{j+1}, D_t^{j+1} with $r < s < t$ such that $D_s^{j+1} \subset D_{h+1}^j$ and $D_r^{j+1}, D_t^{j+1} \subset D_{k-1}^j$. Thus D_s^{j+1} contains no point of K but both D_r^{j+1} and D_t^{j+1} must contain points of K. Since $r < s < t$, K cannot be connected. Thus Y is indecomposable.

131. Miller's Biconnected Set

Let C be a nowhere dense perfect set contained in the unit interval I, and let $W = C \times I \subset R^2$. Let K be an indecomposable continuum such that $K \cap I^2 = W$. The space X is defined using the axiom of choice as follows. Let $\mathfrak{C}$ be the set of composants of K, $\mathfrak{B}$ the set of continua which separate K, and $\mathfrak{D}$ the set of subsets of a fixed countable dense subset Δ of K which are themselves dense in the interior of some square region with edges parallel to I^2 which intersects W. Let $C_1, C_2, C_3, \ldots, C_\alpha, \ldots$ be a well ordering of the elements of $\mathfrak{C}$ where the α are ordinals less than the first ordinal Ω of cardinality c. Likewise let $B_1, \ldots, B_\alpha, \ldots$ and $D_1, \ldots, D_\alpha, \ldots$ be similar well orderings of $\mathfrak{B}$ and $\mathfrak{D}$, respectively. For each $\alpha < \Omega$ define $M_\alpha \subset K$ and a simple closed curve J_α such that

(i) $M_\alpha = p_\alpha \in B_\alpha \cap K$ if $B_\alpha \cap \Delta = \varnothing$;
(ii) $M_\alpha = \varnothing$ if $B_\alpha \cap \Delta \neq \varnothing$;
(iii) For ordinals $\mu \neq \lambda$ and $M_\mu, M_\lambda \neq \varnothing$, M_μ and M_λ belong to different composants of K;

(iv) J_α separates K;
(v) $J_\alpha \cap (\Delta - D_\alpha) = J_\alpha \cap M = \emptyset$ where $M = \bigcup_{\alpha < \Omega} M_\alpha$.

We define the space X to be the set $\Delta \cup M$ with the induced topology from R^2.

1. To show that X is connected, we first note that X is a subset of K. Now if U and V are open sets in R^2 such that $X \subset U \cup V$ and $X \cap U$ and $X \cap V$ are disjoint, then $X \cap U$ and $X \cap V$ are separated sets in R^2 which is completely normal. Thus there exist disjoint open sets U' and V' which separate X. So the complement of $U' \cup V'$ contains a continuum which separates K. Thus by (i) we have a contradiction to $X \subset U \cup V$. So X is connected.

2. Now by (i) and (ii) M_α is either empty or a singleton. Since Δ is countable, (iii) implies that no composant of K contains a connected subset of M for any such set is uncountable. Now if N is a connected subset of X such that $\bar{N}$ is a proper subset of K then $\bar{N}$ is a proper subcontinuum of K and hence lies in some composant so N also lies in the same composant, a contradiction. So the closure of N in X is X, which means that every connected subset of X is dense in X.

3. Consider the family of sets $X \cap B_\alpha$. We now show that there is no set containing at least one element of each $X \cap B_\alpha$ which does not contain all of $X \cap B_\beta$ for some β. Let Q be any set which contains a point of each $X \cap B_\alpha$. Then since X is dense in K, (iv) implies that Q has a point in every set $X \cap J_\alpha$, for J_α is a continuum which separates K and thus any dense subset of K. By (v) and the choice of the D_α, $Q^* = Q \cap (\bigcup_\alpha (X \cap J_\alpha))$ is dense in W. In fact since $J_\alpha \cap M = \emptyset$ we know $Q^* \subset \Delta$ so $J_\alpha \cap D_\alpha \neq \emptyset$ for every α, which implies that every neighborhood of a point of W intersects Q^*. Since $Q^* \subset \Delta$ and is dense in W for some β, $Q^* \supset \Delta_\beta$. But by (v) $X \cap J_\beta = \Delta_\beta \cap J_\beta$ so Q^* and thus Q contains $X \cap J_\beta$.

4. By using the preceding results, we can now show that X is biconnected and contains no dispersion point. If X were the union of two disjoint connected sets X_1 and X_2, X_1 would be dense in X since it is a connected subset of X. Hence X_1 intersects $X \cap B_\alpha$ for every α; since any set with a point in each $X \cap B_\alpha$ contains some $X \cap B_\beta$, X_1 must contain some particular $X \cap B_\beta$. But X_2 must also have a point in every $X \cap B_\alpha$ thus in particular in

$X \cap B_\beta$ so $X_1 \cap X_2 \neq \emptyset$. Thus X is biconnected. Now if $X - p$ is disconnected for any point p, say $X - p = Y_1 \cup Y_2$ where $Y_1 \cap Y_2 = \emptyset$ is a separation, then $Y_1 \cup \{p\}$ is connected but not dense, a contradiction. So X can not have a dispersion point.

132. Wheel without Its Hub

Let X be the closed unit disc in R^2 minus the origin. The topology τ for X is generated by adding to the induced Euclidean topology all open intervals on the radii contained in the open unit disc.

1. X is connected because it is the union of radii homeomorphic to $(0,1]$ all of which intersect the connected unit circle.

2. Furthermore, X is arc connected for we can obtain an arc between two points by running out along the radius containing the first point, along the circumference to the radius containing the second point, and in along this radius to the second point.

3. The Euclidean length of the shortest arc between a and b can be used to define a metric on X; by convention we declare $d(a,a) = 0$. The topology determined by the metric d is precisely τ, and d is a bounded metric with bound $2 + \pi$.

4. As in the order topology, every point is a cut point, for if p is any point of X, then the open radial segment connecting the origin to p and its complement give a separation of $X - \{p\}$.

5. But the topology on X is not the order topology for any linear order on X. Assume that it is, and suppose $a < b < c$ where a,b,c belong to the circumference C. Then since C contains no neighborhood of any of its points, there exists a point $x \in X - C$, such that $a < x < c$. Then $\{t \in C | t < x\}$ and $\{t \in C | t > x\}$ separate C, which is impossible since the circumference C is connected.

133. Tangora's Connected Space

Let X,Y,Z be mutually disjoint and exhaustive dense subsets of the real line R, for example, the diadic rationals (those of the form $m/2^n$), the remaining rationals, and the irrationals. We expand the Euclidean topology on R by adding as open sets X,Y, and sets of the form $\{z\} \cup \{w \in X \cup Y \,\big|\, |w - z| < \delta\}$ where $z \in Z$ and $\delta > 0$.

1. As subspaces, X and Y are totally disconnected since their topology is an expansion of the Euclidean topology in which such sets are totally disconnected. Z is clearly a discrete subspace.

2. $A = Y \cup Z$ and $B = X \cup Z$ are closed being complements of X and Y respectively. Moreover, every point $x \in X$ gives a separation, $\{a \in A | a < x\}$ and $\{a \in A | a > x\}$, of A. Likewise points of Y give separations of B. So A and B are totally disconnected.

3. The space $R = A \cup B$ is connected in the above topology although it is a union of two closed totally disconnected subsets. For suppose C,D is a separation of R where some point of C is less than some point $d \in D$; let $p =$ l.u.b. $\{c \in C | c < d\}$. If p is in X or Y we obtain a contradiction since C and D are both open and hence whichever one contains p contains an open interval of X or Y about p. Likewise, if p $\in Z$, whichever one of C or D contains p contains an open interval in $X \cup Y$ about p, again a contradiction. Hence $A \cup B$ is connected.

134. Bounded Metrics

If (X,d) is any metric space we define new metrics for X by $\delta = d/(1 + d)$ and $\Delta = \min(d,1)$.

1. That δ is indeed a metric follows from the following proof of the triangle inequality:

$$\frac{d(x,y)}{1 + d(x,y)} + \frac{d(y,z)}{1 + d(y,z)} \geq \frac{d(x,y) + d(y,z)}{1 + d(x,y) + d(y,z)}$$
$$= [[d(x,y) + d(y,z)]^{-1} + 1]^{-1} \geq [[d(x,z)]^{-1} + 1]^{-1} = \delta(x,z).$$

2. Since $d = \delta/(1 - \delta)$, the metric δ is equivalent to the original metric d, for any open ball in one metric contains an open ball in the other metric.

3. The metric δ is bounded (by 1), but the space (X,δ) need not be totally bounded. Suppose, for instance, that (X,d) is the real line with the Euclidean metric. Then $B_\delta(x;\epsilon) = \{y | \delta(x,y) < \epsilon\} = \{y | \, |x - y| < \epsilon/(1 - \epsilon)\} = B_d(x;\epsilon/(1 - \epsilon))$. Clearly for small ϵ, no finite number of such balls can cover the real line.

4. By iterating the process by which we derived δ from d, we can produce a sequence of equivalent, bounded metrics $\{d_n\}$, related by $d_{n+1} = d_n/(1 + d_n)$.

5. Δ is also bounded by 1, and is a metric for X since $\min(d(x,y),1)$

$\leq \min(d(x,z) + d(z,y),1) \leq \min(d(x,z),1) + \min(d(z,y),1)$. Clearly Δ is equivalent to d since they agree for all small radii.

6. If (X,d) is the real line with the Euclidean metric, the open ball $B_\Delta(0,1)$ is the interval $(-1,1)$. Its closure $\overline{B_\Delta(0,1)}$ is $[-1,1]$, but the closed ball $\{x|\Delta(0,x) \leq 1\}$ equals X.

7. Any bounded metric on a topological space X can be used to define the Fréchet metric on the product space $X^Z = \prod_{i=1}^{\infty} X_i$ (where each $X_i = X$) which yields the Tychonoff product topology. If δ is the bounded metric on X, we simply define the product metric to be $d^*(x,y) = d^*(\langle x_1,x_2, \ldots\rangle, \langle y_1,y_2, \ldots\rangle) = \Sigma 2^{-i}\delta(x_i,y_i)$. The topology of (X^Z,d^*) can be shown to be the Tychonoff product topology by a direct comparison of basis neighborhoods.

8. If (X,d) is the real line with the Euclidean metric, we can define a special metric by

$$\sigma(x,y) = \left| \frac{x}{1+|x|} - \frac{y}{1+|y|} \right|.$$

Heroic but straightforward calculations can be used to verify that σ is indeed a metric on X, and that it yields the Euclidean topology. In fact, $\sigma(x,y) < |x - y|$ for all $x,y \in X$. But in (X,σ), the positive integers form a Cauchy sequence since $\sigma(n,m) = |n - m|/(1 + |n|)(1 + |m|)$. Of course this Cauchy sequence has no limit point in X, so (X,σ) is not a complete metric space.

135. Sierpinski's Metric Space

If $X = \{x_i|i = 1, 2, 3, \ldots\}$ is a countable set, the function $d(x_i,x_j) = 1 + 1/(i + j)$ for $i \neq j$, $d(x_i,x_i) = 0$ is a metric on X.

1. Since $\{y \in X|d(x_i,y) < \frac{1}{2}\} = \{x_i\}$, each point in X is open. So the topology generated on X by the metric d is the discrete topology.

2. (X,d) is a complete metric space, since all Cauchy sequences are eventually constant.

3. Let $S_n = \{y \in X|d(y,x_n) \leq 1 + 1/2n\}$. Then $S_n = \{x_n, x_{n+1}, x_{n+2}, \ldots\}$, so $\{S_n\}$ is a nested sequence of closed balls whose intersection, $\bigcap S_n$, is empty. Of course the radii of the sets S_n converge to 1, not to 0.

136. Duncan's Space

Let X be the set of strictly increasing infinite sequences of positive integers such that $\delta(\langle x_i \rangle) = \lim_{n \to \infty} N(n,x)/n$ exists, where $N(n,x)$ is the number of elements in the sequence $x = \langle x_i \rangle$ which are less than n. $\delta(x)$ will be called the density of the point $x \in X$. We then define a metric on X by the condition $d(x,y) = k(x,y)^{-1} + |\delta(x) - \delta(y)|$ where $k(x,y)$ is the least integer n for which $x_n \neq y_n$; if $x = y$, we set $d(x,y) = 0$.

1. To verify that (X,d) is a metric space, we need only check that $d(x,y) \leq d(x,z) + d(z,y)$. Certainly $k(x,y) \geq \min\{k(x,z), k(z,y)\}$, and the triangle inequality for real numbers shows that $|\delta(x) - \delta(y)| \leq |\delta(x) - \delta(z)| + |\delta(z) - \delta(y)|$. The triangle inequality for d now follows trivially. (Note that d is a bounded metric since $k(x,y) \geq 1$, and $0 \leq \delta(a) \leq 1$; thus $d(x,y) \leq 2$.)

2. X is a subset of Z^Z, but the topology τ on X is not the same as the subspace topology σ induced from Z^Z. But since the Baire metric $\rho(x,y) = 1/k(x,y)$ yields the topology σ, and since $\rho(x,y) \leq d(x,y)$, we have $\sigma \subset \tau$. Thus the topology on X is an expansion of the induced product topology.

3. Each projection map $\pi_n: X \to Z$ is continuous on (X,σ), therefore also on (X,τ). So any compact subset of X must project to finite subsets since only finite subsets are compact in Z. But every open metric ball B will have some $\pi_n(B)$ which is infinite, for sufficiently large n. This can be seen by observing that $d(x,y) < \epsilon$ means that the sequences x and y agree in their initial terms, and ultimately have densities which are approximately equal. Thus no open subset of (X,τ) can be contained in a compact set, so X is neither locally compact nor compact.

4. Suppose $X = \bigcup_{k=1}^{\infty} Y_k$ where each Y_k is compact. Let $m(n,k)$ be the greatest integer in the finite set $\pi_n(Y_k)$. Certainly there exists in X a point $\langle x_i \rangle$ where for each i, $x_i > m(i,i)$; such a point would be in none of the sets Y_k, thus our supposition is untenable. Hence X is not σ-compact.

5. The set of all finite sequences of integers is countable, as is the set of all arithmetic sequences. Thus the set of all $x \in X$ of the form $\langle x_1, x_2, \ldots, x_n, x_{n+1}, \ldots \rangle$ where $\langle x_{n+1}, x_{n+2}, \ldots \rangle$ is an arithmetic sequence is countable. Furthermore, the collection of

all finite unions of such sequences is clearly dense in X, so X is separable.

6. Since Z^Z is totally separated, so is (X,σ) and therefore also (X,τ). But no point in (X,τ) is isolated, so X is dense-in-itself and thus not scattered. Since X is metrizable but not discrete, it is not extremally disconnected.

137. Cauchy Completion

If (X,d) is a metric space, we let X^* be the set of all equivalence classes of Cauchy sequences where the sequence $\{x_n\}$ is equivalent to $\{y_n\}$ iff $\lim_{n\to\infty} d(x_n,y_n) = 0$. We define on X^* a metric d^* by $d^*(x^*,y^*) = \lim_{n\to\infty} d(x_n,y_n)$, where $\{x_n\}$ is any element in the equivalence class x^*, and similarly, $\{y_n\} \in y^*$.

1. The metric space (X^*,d^*) is complete, for if $\{x_n{}^*\}$ is a Cauchy sequence in X^*, and if for each n, $\{x_{i,n}\}$ is a representative sequence in the equivalence class $x_n{}^*$, then a receding diagonal sequence $\{x_{i,n_i}\}$ is a Cauchy sequence, provided n_i is chosen so that $|x_{i,n_i} - x_{i,m}| < 2^{-i}$ for all $m > n_i$. The equivalence class of this sequence is the limit of the sequence $\{x_n{}^*\}$.

2. The mapping $f\colon X \to X^*$ which takes each point $x \in X$ into the equivalence class containing the constant sequence $\{x, x, x, \ldots\}$ is a distance preserving injection of X into X^*. That is, $d^*(f(x),f(y)) = d(x,y)$. $f(X)$, the image of f in X^*, is a dense subset of X^*.

138. Hausdorff's Metric Topology

Let (S,d) be a metric space, and let X be the collection of all nonempty bounded closed subsets of S. Let $f: S \times X \to R^+$ be defined by $f(s,B) = \inf_{b\in B} d(s,b)$, and let $g\colon X \times X \to R^+$ be given by $g(A,B) = \sup_{a\in A} f(a,B)$, and let $\delta(A,B) = \max\{g(A,B),g(B,A)\}$. (X,δ) is known as Hausdorff's metric space.

1. If for some $s \in S$ and $B \in X$, $f(s,B) = 0$, we must have $s \in B$ since s would be a limit point of B. Thus if $\delta(A,B) = 0$, every point of A must belong to B and conversely; thus in this case, $A = B$.

2. Since $d(a,b) \leq d(a,c) + d(c,b)$ for all a, b, $c \in S$, we have: $\inf_B d(a,b) \leq d(a,c) + \inf_B d(c,b)$. Hence $f(a,B) \leq \inf_C d(a,c) +$

$\inf_C f(c,B)$, so $f(a,B) \leq f(a,C) + \sup_C f(c,B)$. This then yields $\sup_A f(a,B) \leq \sup_A f(a,C) + g(C,B)$, or $g(A,B) \leq g(A,C) + g(C,B)$. Thus δ satisfies the triangle inequality and is thus a metric for X.

139. The Post Office Metric

Let (X,d) be the Euclidean plane with the ordinary metric; let 0 be the origin in this plane. We define a new metric d^* on X by the formula $d^*(p,q) = d(0,p) + d(0,q)$, whenever $p,q \in X$, $p \neq q$; if $p = q$, we let $d^*(p,q) = 0$.

1. d^* is a metric for X since clearly $d^*(p,q) = d(0,p) + d(0,q) \leq d(0,r) + d(0,p) + d(0,r) + d(0,q) = d^*(p,r) + d^*(r,q)$.
2. Every point but 0 is open since if $p \neq 0$, and if $\epsilon = \frac{1}{2}d(0,p)$ the open metric ball around p of radius ϵ is just $\{p\}$. Basis neighborhoods of 0 are just Euclidean open balls.
3. Since each point of $X - \{0\}$ is open, X is not separable and thus neither σ-compact nor compact. Although each point of $X - \{0\}$ has a compact neighborhood (namely, itself) 0 does not; thus X is not locally compact.
4. The open metric balls around 0 are also closed, so X is zero dimensional. However X is not extremally disconnected since the closure of the open set $E = \{(x,y) | y > 0\}$ is $E \cup \{0\}$, which is not open. But X is scattered, since every nonempty subset contains isolated points.

140. The Radial Metric

Let (X,d) be the Euclidean plane with the ordinary metric; let 0 be the origin in this plane. We define a new metric d^* on X by the composite formula:

$$d^*(p,q) = \begin{cases} 0 & \text{if } p = q \\ d(p,q) & \text{if } p \neq q \text{ and the line through } p \text{ and } q \text{ passes} \\ & \text{through the origin.} \\ d(p,0) + d(q,0) & \text{otherwise.} \end{cases}$$

The metric d^* corresponds to a model in which all distances are measured along lines radiating from the origin.

1. The metric balls around points removed from the origin consist simply of line segments lying on a radial path through the point;

in addition, points near 0 have a Euclidean neighborhood of 0 included in their metric balls.

2. (X,d^*) is not separable, for the closure of any subset A can include only points which lie on rays from the origin which pass through points of A. For a countable subset A, only countably many such rays can intersect $\bar{A}$, so $\bar{A} \neq X$.

3. The induced topology on each ray through the origin is the Euclidean topology. Thus X is arc connected since the rays which connect points a and b to the origin are arcs.

141. Radial Interval Topology

We generate a topology τ on the coordinate plane X from a basis consisting of all open intervals disjoint from the origin which lie on lines passing through the origin, together with sets of the form $\bigcup\{I_\theta | 0 \leq \theta < \pi\}$ where each I_θ is a nonempty open interval centered at the origin on the line of slope $\tan \theta$.

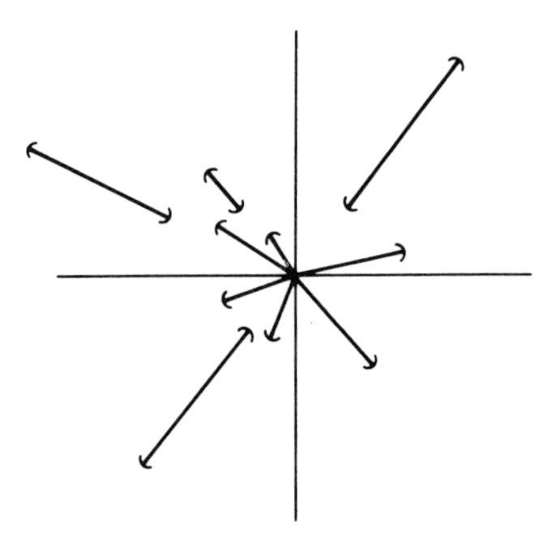

1. The topology τ is strictly finer than the topology σ given by the radial metric on X since every metric ball is in τ, but not every τ-neighborhood of the origin is a union of metric balls.

2. (X,τ) is, like the radial metric topology, completely normal. For it is clearly Hausdorff, and suppose A and B were separated subsets of X. Then if L_θ is the line of slope $\tan \theta$ passing through the origin, $A \cap L_\theta$ and $B \cap L_\theta$ are separated subsets of L_θ which as a subspace carries the Euclidean topology. Thus there exist in L_θ disjoint Euclidean open sets U_θ and V_θ containing $A \cap L_\theta$ and $B \cap L_\theta$; clearly $\bigcup U_\theta$ and $\bigcup V_\theta$ are disjoint open subsets of X containing A and B.

3. (X,τ) is not first countable because it does not have a countable local basis at the origin. For suppose $\{U_n\}$ were such a basis where $U_n = \bigcup I_\theta^n$. If $\{\theta_n\}$ is any sequence of angles, and if J_{θ_n} is the middle half of the interval $I_{\theta_n}^n$, and otherwise $J_\theta = L_\theta$,

then $\cup J_\theta$ is a neighborhood of the origin which contains no set U_n.

4. Clearly (X,τ) is neither Lindelöf nor locally compact: no neighborhood of the origin has a compact closure. Similarly, X is not separable.

5. X is paracompact since each subspace L_θ is paracompact. If $\{U_\alpha\}$ covers X, $\{U_\alpha \cap L_\theta\}$ has a locally finite refinement N_α^θ (in which for each θ, a neighborhood of the origin 0 belongs to at most one of the sets N_α^θ) so the collection $\{N_\alpha^\theta | 0 \notin N_\alpha^\theta\}$ together with $\cup\{N_\alpha^\theta | 0 \in N_\alpha^\theta\}$ is a locally finite refinement of $\{U_\alpha\}$.

6. X is not metrizable since it is not first countable. Thus, since it is regular, it cannot have a σ-locally finite base. This can be seen directly by assuming $\bigcup_{i=1}^{\infty} B_i$ is a base for τ, where each B_i is a locally finite family. This would mean that $\{U \in \bigcup_{i=1}^{\infty} B_i | 0 \in U\}$ is a countable local basis at 0, which does not exist.

142. Bing's Discrete Extension Space

143. Michael's Closed Subspace

If R is the set of real numbers with power set P, we let $X = \prod_{\lambda \in P} \{0,1\}_\lambda$ where $\{0,1\}_\lambda$ is a copy of the two point discrete space. For each $r \in R$, let x_r be the point of X whose λth coordinate $(x_r)_\lambda$ equals 1 iff $r \in \lambda$; let $M = \{x_r \in X | r \in R\}$. (If we think of X as the power set of P, M becomes the collection of principal ultrafilters of R.) Now if X has the Tychonoff topology τ, $X - M$ is clearly dense in X, so we may form the discrete extension σ of τ by $X - M$. In (X,σ), each point of $X - M$ is open, while each point of M retains its τ-neighborhoods. Let Y be the subspace $M \cup F$ of (X,σ) where F is the collection of all x in $X - M$ such that $(x)_\lambda = 1$ for only finitely many λ.

1. Since $X - M$ is open in (X,σ), M is a closed subset of (X,σ). As a subspace, M inherits from τ, and thus from σ, the discrete topology, since if $x_r \in M$, and if $\lambda = \{r\}$, then $\pi_\lambda^{-1}(1) \cap M = \{x_r\}$. A slight extension of this argument shows that any two disjoint subsets of M are contained in disjoint open subsets of X: if λ is a subset of R corresponding to a subset L of M, then

$\pi_\lambda^{-1}(1)$ and $\pi_\lambda^{-1}(0)$ are disjoint open sets in X which contain L and $M - L$, respectively.

2. Now (X,σ) is clearly Hausdorff, so to show that it is normal we need only find disjoint open neighborhoods for disjoint closed sets A_1 and A_2. Let U_1 and U_2 be disjoint open sets in X which contain, respectively, $A_1 \cap M$ and $A_2 \cap M$. Then $(U_1 - A_2) \cup (A_1 - M)$ and $(U_2 - A_1) \cup (A_2 - M)$ are disjoint open sets containing A_1 and A_2, respectively.

3. Since $X - Y \subset X - M$, Y is closed in (X,σ). Thus, since (X,σ) is normal, so is the subspace Y.

4. Neither (X,σ) nor Y is paracompact, since the covering by basis sets has no locally finite refinement, for every neighborhood of a point $x_r \in M$ must contain infinitely many points of $X - M$ (or of F).

5. (X,σ) is not even metacompact, since some points of $X - M$ will always lie in infinitely many neighborhoods of points of M, so the covering by basis sets has no point finite refinement. But Y, which does not contain all the points of $X - M$, is metacompact. For let $\{U_\alpha\}$ cover Y; select for each $x_r \in M$ a neighborhood $U_r \in \{U_\alpha\}$. Then $V_r = \{x \in U_r | \{r\} \in x\} = U_r \cap \pi_{\{r\}}^{-1}(1)$ is open in Y, and $\{V_r\}$ is point finite. Thus the family $\{V_r\}$ together with the singletons of $F = Y - M$ form a point finite refinement of $\{U_\alpha\}$.

6. Since Y is normal and metacompact, it is countably paracompact; but it is not paracompact.

PART III
Metrization Theory

Conjectures and Counterexamples

PROLOGUE

The search for necessary and sufficient conditions for the metrizability of topological spaces is one of the oldest and most productive problems of point set topology. Alexandroff and Urysohn [6] provided one solution as early as 1923 by imposing special conditions on a sequence of open coverings. Nearly ten years later R. L. Moore chose to begin his classic text on the Foundations of Point Set Theory [82] with an axiom structure which was a slight variation of the Alexandroff and Urysohn metrizability conditions. After Jones [56], we now call any space which satisfies Axiom 0 and parts 1, 2, 3 of Axiom 1 of [82] a Moore space. Each metric space is a Moore space, but not conversely, so the search for a metrization theorem became that of determining precisely which Moore spaces are metrizable. The most famous conjecture was that each normal Moore space is metrizable.

It would probably be no exaggeration to say that for the last 30 years, the normal Moore space conjecture dominated the search for a significant metrization theorem and in the process played a major role in the development of point set topology. The conjecture itself was first stated in 1937 by Jones [56] who showed that if $2^{\aleph_0} < 2^{\aleph_1}$, then every separable normal Moore space is metrizable. The next major result came nearly twenty years later when Bing [22] and Nagami [86] showed that every paracompact Moore space is metrizable. But Jones' result together with more recent ones of Heath [50] and Bing [20] indicated a close relationship between the normal Moore space conjecture and the continuum hypothesis which was shown by Cohen [32] in 1963 to be independent of the axioms of set theory. Recently Silver (see Tall [118]) used a Cohen model to show that the normal Moore space conjecture itself could not be proved from the present axioms of set theory.

Thus as metrization research shifts from topology to set theory, we survey in this paper the chief topological milestones of the last half century. We shall not present proofs that are available in the literature, but shall concentrate instead on gathering together the most significant definitions, theorems, conjectures and counterexamples. The latter will be grouped together at the end of the paper and referenced throughout the text whenever appropriate. We begin at the beginning.

Basic Definitions

We shall assume throughout this paper that all topological spaces are Hausdorff. Most often we shall be concerned only with regular spaces, though we do not formally require this assumption. **Regular** spaces are those which admit a separation of a point from a closed set by disjoint open neighborhoods. A space X is **normal** if each pair of disjoint closed sets can be separated by disjoint open neighborhoods, and **completely normal** if the same can be done for separated sets. A space is completely normal if and only if it is **hereditarily normal** [42], that is, if and only if every subspace is normal.

A subset of a topological space which can be written as the countable union of closed sets is called an ***F_σ*-set;** the complement of an F_σ-set can be written as a countable intersection of open sets, and is called a ***G_δ*-set** (or an **inner limiting** set). A space in which every closed set is G_δ (or equivalently, every open set is F_σ) will be called a ***G_δ*-space;** a normal space which is also a G_δ-space is called (by Čech [29]) **perfectly normal.** Every metric space is perfectly normal and every perfectly normal space is completely normal [68], so we have the following implications:

$$\text{Metrizable} \Rightarrow \begin{matrix}\text{perfectly}\\ \text{normal}\end{matrix} \Rightarrow \begin{matrix}\text{completely}\\ \text{normal}\end{matrix} \Rightarrow \text{normal} \Rightarrow \text{regular.}$$

Examples 5, 2, 10, and 6 below show that none of these implications is reversible.

If a topological space has a countable dense subset it is called **separable,** if it has a countable basis it is **perfectly separable** (or **second countable**), and it it has a countable local basis at each point it is **first countable.** A space in which every subspace is separable is called **hereditarily separable.** If every open covering of X has a countable subcovering, X is called **Lindelöf** (or, by Russian mathematicians, **finally compact** [3]); clearly each perfectly separable space is both Lindelöf and hereditarily separable.

Since in a metric space the (open) balls of radius $1/n$ form a countable local basis at each point, every metric space is first countable. Metric spaces need not be second countable, but in metric spaces the properties of separable, hereditarily separable, second countable and Lindelöf coincide. Urysohn [129] proved in 1925 that every normal second countable space is metrizable, and, in response to a question proposed by Urysohn, Tychonoff [126] showed a year later that every regular second countable space is metrizable.

Developments

A collection of sets $F = \{U_\alpha\}$ is said to **cover** a space X if each point of X belongs to some U_α; if each U_α is open, the cover F is called an **open covering** of X. A cover $\{V_\beta\}$ of a space X is a **refinement** of a cover $\{U_\alpha\}$ if for each V_β there is a U_α such that $U_\alpha \subset V_\beta$. If $S \subset X$, the **star** of S with respect to a cover $F = \{U_\alpha\}$ is the union of all sets in F which intersect S; the star of S is denoted by $F^*(S)$, and the star of the singleton $\{x\}$ is usually denoted simply by $F^*(x)$.

A **development** for a topological space X is a countable family $\mathfrak{F}$ of open coverings F_i such that if C is a closed subset of X and $p \in X - C$, there is a covering $F \in \mathfrak{F}$ such that no element of F which contains p intersects C (i.e., such that $F^*(p) \cap C = \varnothing$). A space with a development is called **developable.** If $\mathfrak{F} = \{F_i\}$ is a development where $F_i \subset F_{i+1}$ for all i, the family $\mathfrak{F}$ is called a **nested development,** and if F_{i+1} is a refinement of F_i, $\mathfrak{F}$ is called a **refined development.** Clearly each nested development is a refined development; Vickery [133] showed that every developable space has a nested development. Axiom 0 and parts 1, 2, and 3 of Axiom 1 of Moore [82] require precisely that a space be regular with a nested development $\{F_i\}$; such spaces are called **Moore** spaces (after Jones [56]), and are characterized by the fact that for each $p \in X$, $\{F_i^*(p)\}$ is a countable local basis. Vickery's theorem can be restated as follows: a topological space is a Moore space if and only if it is regular and developable.

Each metric space is a Moore space since the sequence of open coverings by metric balls of radius $1/n$ is a development; Examples 6, 9, 14, and 15 below show that Moore spaces need not be metrizable. In [132] van Douwen gives a generalized version of a construction of Pixley and Roy [93] which yields many simple examples of bad (very nonmetrizable) Moore spaces.

Semimetric Spaces

A **semimetric** for a Hausdorff space X is a symmetric function $d : X \times X \to R^+$ such that $d(x,y) = 0$ if and only if $x = y$, and if $x \in X$

and $E \subset X$, inf $\{d(x,y) \mid y \in E\} = 0$ if and only if $x \in \bar{E}$, the closure of E; a Hausdorff space which admits a semimetric is called a **semimetric space.** If we did not require d to be symmetric, to assert the existence of a function with the remaining properties would be equivalent to saying that the space X was first countable [26]. Thus a semimetric space may be thought of as a first countable space with a symmetric function d. In fact, some Russian mathematicians call these spaces **symmetrizable.**

Now every developable space has a natural semimetric: if $\{F_n\}$ is a nested development for X (with $X \in F_1$), we define $d(x,y) = \inf \{1/n \mid x, y \in U \in F_n\}$. Then d is a semimetric, but clearly not a metric since d is not continuous. Semimetric spaces share with metric spaces the property that every closed set is a G_δ [72], hence such spaces are G_δ-spaces. We use Figure 13 to summarize the implications for regular spaces; counterexamples to the converse implications are listed below each implication.

Every known example (in standard set theory) of a Moore space which

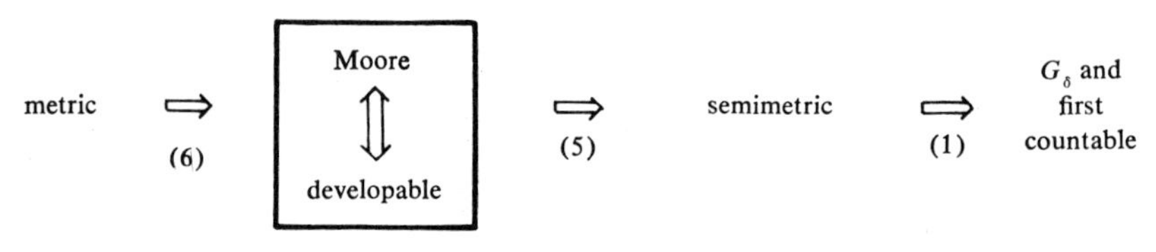

Figure 13.

is not metrizable is also not normal; the normal Moore space conjecture asserts that it will always be thus. Jones [56] in 1937 mounted the first major attack on this conjecture, and succeeded only in proving several weaker theorems: every normal Moore space is completely normal, and every separable normal Moore space is metrizable provided $2^{\aleph_1} > 2^{\aleph_0}$ — a fact implied by (but not equivalent to) the continuum hypothesis. Both of Jones' results have recently been strengthened: McAuley [73] observed in 1954 that a simple modification of Jones' proof will show that every normal semimetric space is completely normal, while in 1964 Heath [49] showed that a necessary and sufficient condition for the metrizability of a separable Moore space is that every uncountable subset M of the real line contains a subset which is not F_σ (in M). This condition is (perhaps not strictly [49]) weaker than that used by Jones, namely $2^{\aleph_0} < 2^{\aleph_1}$.

Jones actually showed that if $2^{\aleph_0} < 2^{\aleph_1}$, then every separable normal space has the property that every uncountable subset has a limit point;

Heath [50] called spaces with this property **$\aleph_1$-compact** and proved the converse to Jones' theorem: if every separable normal space is $\aleph_1$-compact, then $2^{\aleph_0} < 2^{\aleph_1}$.

Paracompactness

The most significant general approximation to the normal Moore space conjecture is the Bing-Nagami theorem that every paracompact Moore space is metrizable. To develop the concept of paracompactness and all its variations, we must first discuss the naming of various covers.

A cover is **point finite** if each point belongs to only finitely many sets in F, **locally finite** if each point has some neighborhood which intersects only finitely many members of F, and **star finite** if each set in F intersects only a finite number of other sets in F. A cover $V = \{V_\beta\}$ of X is a **star refinement** (or a **point star refinement,** or a **Δ refinement**) of a cover $\{U_\alpha\}$ if for each $x \in X$ there is some U_α such that $V^*(x) \subset U_\alpha$ (where $V^*(x)$ is the star of x with respect to $V = \{V_\beta\}$).

A Hausdorff space is called **fully normal** if every open cover has an open star refinement, **strongly paracompact** (or **star paracompact**) if every open cover has an open star finite refinement, **paracompact** if every open cover has an open locally finite refinement, and **metacompact** (or **pointwise paracompact,** or **weakly paracompact**) if every open cover has an open point finite refinement.

Fully normal spaces were first defined by Tukey [125] in 1940, while paracompact spaces were introduced by Dieudonné [33] in 1944. Tukey showed that every metrizable space is fully normal, while Dieudonné showed that every paracompact space is normal. The key link between these definitions was provided by Stone [115] in 1948 who showed that every metric space is paracompact by proving that every fully normal space is paracompact, and conversely. Although a regular semimetric space need not be paracompact (Example 6), Ceder [30] showed that each regular hereditarily separable semimetric space is paracompact. Smirnov [110] showed that a paracompact space which fails to be metrizable must fail for local reasons: every locally metrizable paracompact space is metrizable.

Also in 1948 Morita [84] introduced the concept (but not the name) of strongly paracompact spaces; he showed that each regular Lindelöf space is strongly paracompact while every strongly paracompact space is *a fortiori* paracompact. Kaplan [63] and Alexandroff [1] showed that each separable metric space is strongly paracompact, and that a nonseparable metric space need not be strongly paracompact (Example 11 below). We summarize in Figure 14 these results together with the counterexamples to the converse implications.

A most important variation of paracompact spaces is that of **countably paracompact** spaces, those for which every countable open covering has a locally finite open refinement. Morita [84] showed in 1948 that every metacompact normal space is countably paracompact, (see also Michael [77]) while in 1951 Dowker [34] proved that every perfectly normal space is countably paracompact. Countably paracompact normal spaces are sometimes called **binormal;** they have been characterized in many ways by Mansfield [69] and Dowker [34]. Clearly every fully normal (i.e., paracompact) space is binormal, and every binormal space is normal.

In [20] Dowker conjectured that every normal space is countably paracompact, and showed that this conjecture is equivalent to the conjecture that the product of a normal space with the closed unit interval I is

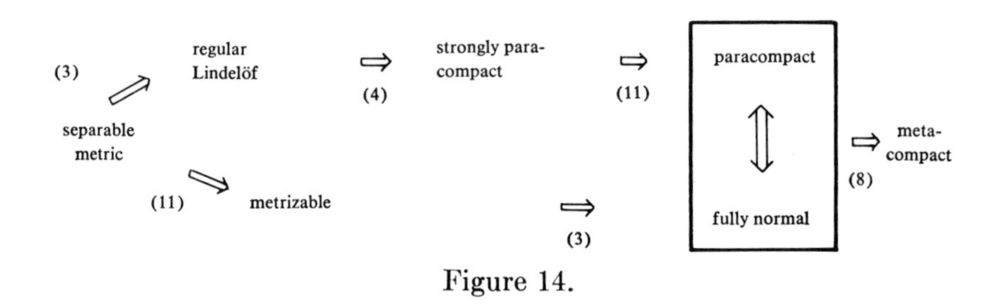

Figure 14.

normal by showing that X is countably paracompact and normal if and only if $X \times I$ is normal. Dowker's conjecture stood for twenty years before it was proved false (see [103]), motivating considerable work on the relation between normality, paracompactness, and product spaces.

Screenable Spaces

A collection $\mathcal{B}$ of sets is called **conservative** (or **closure preserving**) if for every subcollection $\mathcal{A} \subset \mathcal{B}$, the union of the closure of the members of $\mathcal{A}$ is closed. A conservative collection is **discrete** if the closures are pairwise disjoint. Equivalently a collection $\mathcal{B}$ of subsets of X is discrete if every point in X has a neighborhood which intersects at most one of the sets in $\mathcal{B}$.

Now a topological space is called (by Bing [22]) **screenable** if for each open covering F there is a sequence F_n of collections of pairwise disjoint open sets such that $\bigcup F_n$ is a refinement of F. The space is called **strongly screenable** if the F_n may be chosen to be discrete. A **perfectly screenable** space is one with a **σ-discrete base**—that is, a base which is the countable union of discrete families. A formally weaker condition is that of a **σ-locally finite base**—one which is the countable union of locally finite families. It follows directly from the definitions that every perfectly screenable space is strongly screenable, and *a fortiori*, screenable.

Stone [115] showed in 1948 that every metric space has a σ-discrete (and thus σ-locally finite) base. Shortly thereafter, Nagata [87] and Smirnov [110] showed that every regular space with a σ-locally finite base is metrizable, while Bing [22] showed that each perfectly screenable regular

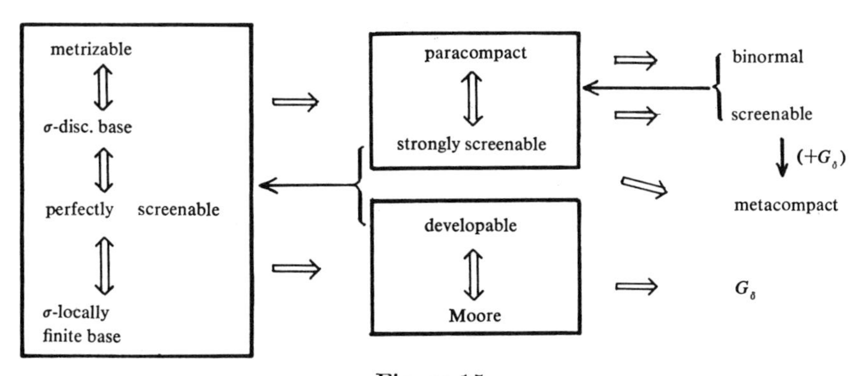

Figure 15.

space is metrizable. A few years after Bing's work appeared, Nagami [86] showed that in regular spaces paracompactness is equivalent to strong screenability and that in binormal (i.e., countably paracompact and normal) spaces, screenable implies strongly screenable. Every strongly screenable developable space must be perfectly screenable since the discrete refinements of the development will form a σ-discrete base [22]. Thus every paracompact Moore space is metrizable, for by Nagami's theorem such spaces are strongly screenable and developable. Heath [49] showed that every screenable G_δ-space (thus every screenable developable space) is metacompact.

We summarize in Figure 15 the major implications for regular spaces (which are really the only ones of interest *vis-à-vis* metrizability). The relevant counterexamples are classified by the Venn diagram in Figure 16.

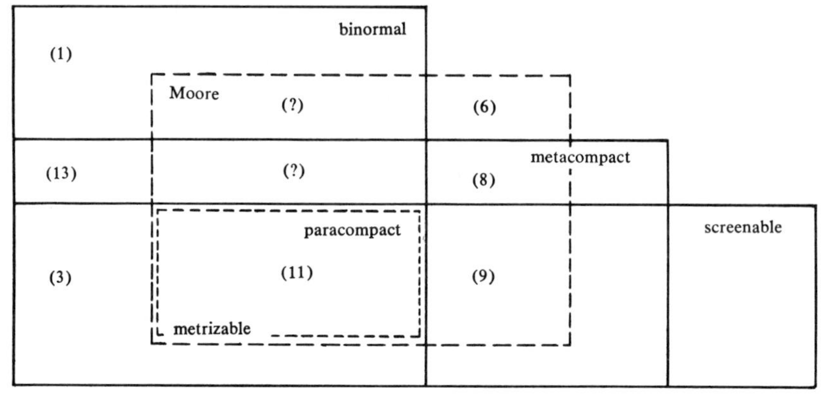

Figure 16.

Collectionwise Normal Spaces

A (Hausdorff) topological space is called **collectionwise normal** if every discrete collection of sets (or, equivalently, closed sets) can be covered by a pairwise disjoint collection of open sets, each of which covers just one of the original sets. If we weaken this property by requiring it of only countable discrete collections, we call the space **countably collectionwise normal.** On the other hand, we may strengthen collectionwise normal by requiring every **almost discrete** collection of sets (that is, a collection which is discrete with respect to its union) to have a covering by pairwise disjoint open sets: such spaces are called **completely collectionwise normal.** A space is completely collectionwise normal if and only if it is hereditarily collectionwise normal [29], so each completely collectionwise normal space must be completely normal (i.e., hereditarily normal). Every metric space is completely collectionwise normal, so we summarize the implications in Figure 17. Examples 10 and 12 below show that normal spaces need not be collectionwise normal, and that collectionwise normal spaces need not be completely collectionwise normal.

Bing [22] showed that every fully normal (i.e., paracompact) space is collectionwise normal; Nagami [86] showed that every metacompact collectionwise normal space is strongly screenable. Nagami and Michael [75] showed that the converse holds for regular spaces. So for regular spaces, the concepts of fully normal, paracompact and strongly screenable coincide. Since each strongly screenable developable space is perfectly screenable and each regular perfect screenable space is metrizable, we conclude again that every paracompact Moore space is metrizable. In fact, Bing [22] gave two slightly stronger results: every screenable, normal Moore space is metrizable (since every screenable normal developable space is strongly screenable) and every collectionwise normal Moore space is metrizable (since every such space is screenable). Thus to prove every normal Moore space metrizable, it would suffice to prove it collectionwise normal. In 1964 Bing [20] showed that every normal Moore space is countably collectionwise normal.

Several conditional converses of the basic implications have been established. Michael [77] showed that every collectionwise normal metacompact space is paracompact, while McAuley [72] showed that every collectionwise normal semimetric space is paracompact, and that every paracompact semimetric space is completely collectionwise normal.

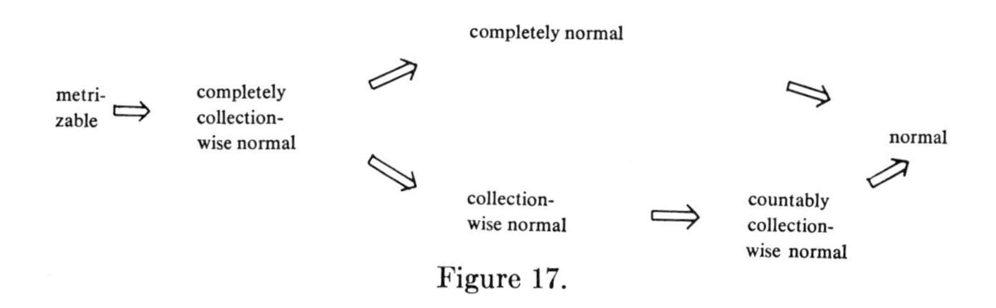

Figure 17.

In 1960 Alexandroff [2] developed a slightly different type of metrization theorem by defining the concept of a uniform base: a basis for X is a **uniform base** if for each $x \in X$ and each neighborhood U of x, only a finite number of the basis sets which contain x intersect $X - U$. Equivalently, a base $\mathcal{B}$ for X is uniform if for each $x \in X$ any infinite subset of $\{U \in \mathcal{B} \mid x \in U\}$ is a (local) basis at x. Since for each integer n the open covering of a metric space by balls of radius $1/n$ has a locally finite subcovering, each metric space has a uniform base, and each space with a uniform base is metacompact. Alexandroff showed that a collectionwise normal space with a uniform base is metrizable, and similarly that a paracompact space with a uniform base is metrizable. Heath [49] proved that a regular space has a uniform base if and only if it is metacompact and developable, from which both of Alexandroff's theorems follow.

Arhangel'skiĭ [12] strengthened the definition of a uniform base by substituting for the point x an arbitrary compact set K: he called $\mathcal{B}$ a **strongly uniform base** if for any compact subset $K \subset X$ and any neighborhood U of K, only a finite number of the basis sets intersect both

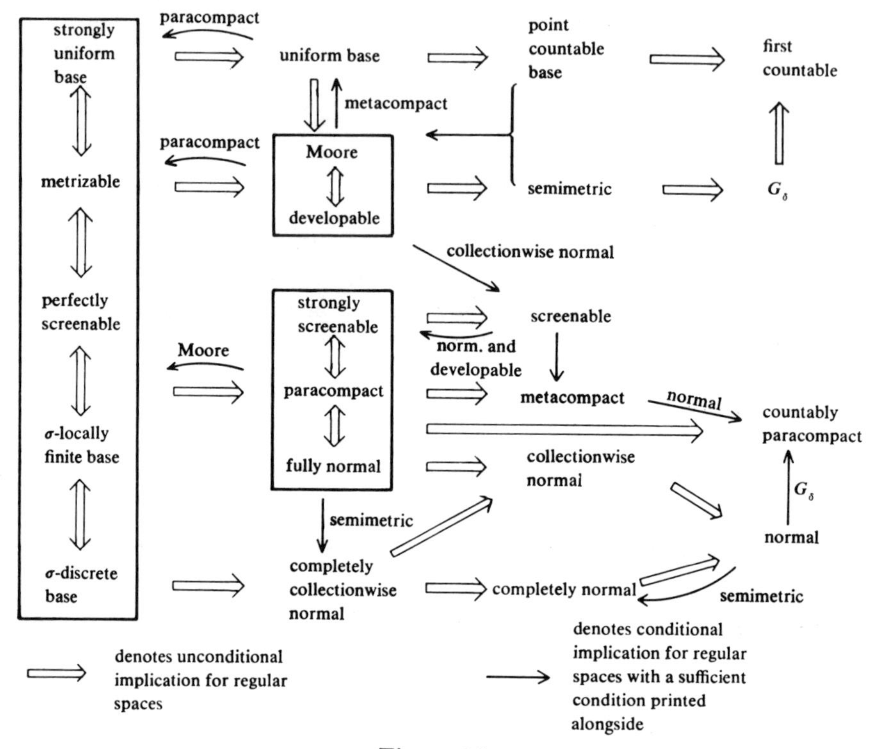

Figure 18.

K and $X - U$. Arhangel'skiĭ showed [14] that a space is metrizable if and only if it has a strongly uniform base. Finally, a space is said to have a **point countable base** if it has a basis $\mathcal{B}$ such that no point is contained in more than countably many sets of $\mathcal{B}$. Each uniform base is point countable, and Heath [48] has shown that every semimetric space with a point countable base is developable. We summarize the preceding implications in Figure 18; the reader is invited to draw the corresponding Venn diagram.

Conjectures

The literature on the normal Moore space conjecture abounds in conditional theorems which assert that if some hypothesis is true, then some particular theorem is true. A famous example cited previously is Jones' theorem that if $2^{\aleph_0} < 2^{\aleph_1}$, then every separable normal Moore space is metrizable. These theorems deal with implications among statements whose truth or falsehood is either not yet known, or which are in some cases (e.g., the continuum hypothesis) independent of the axioms of set theory.

We shall denote by ***CH*** the continuum hypothesis $2^{\aleph_0} = \aleph_1$; Gödel [44] and Cohen [32] proved this hypothesis consistent with and independent of the Zermelo-Fraenkel (or Gödel-Bernays) axioms of set theory (hereafter referred to simply as "set theory"). We shall denote by ***WCH*** Jones' hypothesis that $2^{\aleph_0} < 2^{\aleph_1}$, since it is a weak version of *CH*: if $2^{\aleph_0} = \aleph_1$, then $2^{\aleph_0} = \aleph_1 < 2^{\aleph_1}$ by Cantor's theorem. Clearly the consistency of *CH* implies the consistency of *WCH*. The negation of *WCH*, namely $2^{\aleph_0} = 2^{\aleph_1}$, is called the Luzin Hypothesis ***LH***; Bukovsky [27] showed that *LH* is consistent with set theory. Thus *WCH*, the negation of *LH*, is independent of set theory.

Since every separable metric space has $2^{\aleph_0}$ Borel subsets *WCH* implies that every separable uncountable metric space has a subset which is not a Borel set; we shall call this ***BH***, for Borel hypothesis. Heath [49] used a special case of *BH* to strengthen Jones' theorem: we shall denote by ***HH*** the statement that every uncountable subspace M of the real line contains a subset which is not F_σ in M. Since every F_σ-set is a Borel set, *BH* implies *HH*; Heath showed that *HH* is equivalent to Jones' conjecture ***JC*** that every separable normal Moore space is metrizable. The consistency of the continuum hypothesis implies that of *JC*, while the independence of *JC* was proved by Tall and Silver [118] in 1970.

Heath also showed that Jones' conjecture follows from the hypothesis ***MMSC*** that every normal metacompact Moore space is metrizable; clearly *MMSC* is weaker than the normal Moore space conjecture ***MSC.*** *MMSC* is equivalent to Alexandroff's conjecture ***AC*** that every normal space with a uniform base is metrizable [3]. Traylor [124] suggested the conjecture ***TC*** that every normal Moore space is metacompact. Since McAuley [72] showed that a separable normal metacompact Moore space is metrizable, Traylor's conjecture implies Jones' conjecture.

Several common conjectures center on semimetric spaces, a generalization of Moore spaces. Brown [26] suggested that every normal semimetric space is collectionwise normal, while Heath [47] appeared to strengthen this conjecture by suggesting that every normal semimetric space is paracompact. Actually since every semimetric collectionwise normal space is paracompact [72], these conjectures are equivalent; we shall denote them by ***NSP.*** McAuley [74] proposed the weaker conjecture ***SNSP*** that every separable normal semimetric space is paracompact. The Bing-Nagami result that every paracompact Moore space is metrizable shows that *NSP*

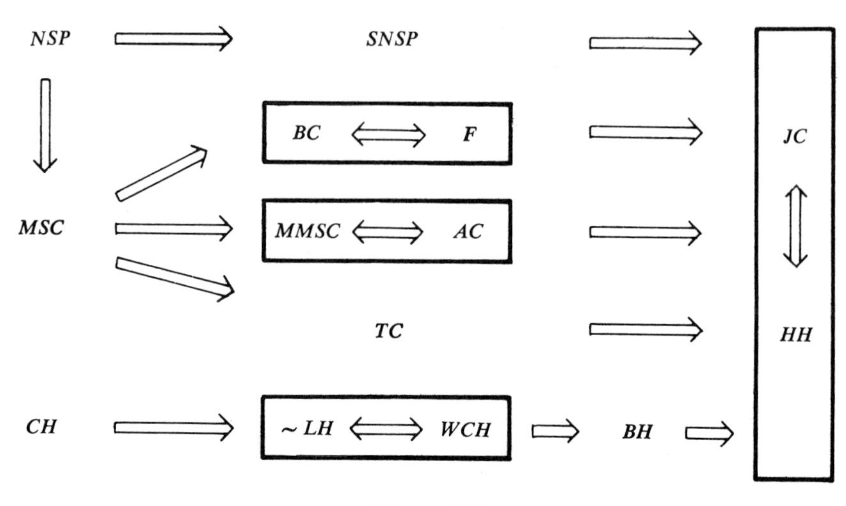

Figure 19.

implies the Moore space conjecture *MSC*, and similarly, *SNSP* implies Jones' separable Moore space conjecture *JC*.

In [22] Bing showed that *MSC* is equivalent to the conjecture that every normal Moore space is collectionwise normal; in [20], he considered the weaker conjecture ***BC*** that every normal Moore space is collectionwise normal with respect to a discrete collection of points. (This property is called, naturally enough, **collectionwise Hausdorff.** Bing termed a counterexample to *BC* one of **type *D*.**) Bing showed that *BC* is equivalent to the following set theoretic conjecture: If X is a set and if Y denotes the product $X \times X$ less the diagonal $\Delta = \{(x,x) \in X \times X\}$, we call a subset $W \subset Y$ a **skew subset** if the projections $\pi_x(W)$ and $\pi_y(W)$ are disjoint. Bing's alternative to *BC* is the conjecture ***F*** that if $f: Y \to Z^+$ is a function from Y to the nonnegative integers with the property that for each skew subset $W \subset Y$ there is a function $F_W: W \to Z^+$ which dominates f in the sense that max $[F_W(x), F_W(y)] > f(x,y)$ for all $(x,y) \in W$, then there is a function $F: X \to Z^+$ which dominates f in this sense for all $(x,y) \in Y$.

Bing also showed that *BC* implies *JC* by showing that any nonmetrizable separable normal Moore space would necessarily be a counterexample of type *D*. We summarize the relationships among these conjectures in Figure 19. Since all of the conjectures in this figure imply *JC*, none of them can be proved from the axioms of set theory. But the consistency of these various hypotheses (except of course for *CH* and its consequences) remains an open question.

We have already mentioned Dowker's conjecture ***DC*** that every nor-

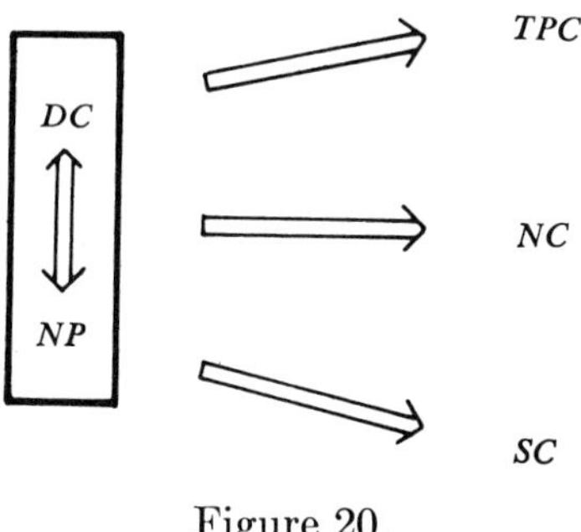

Figure 20.

mal space is countably paracompact; he showed this equivalent to the conjecture ***NP*** that the product of every normal space with the unit interval is normal [34]. Nagami [86] showed that a screenable normal countably paracompact space is paracompact and conjectured ***NC*** that every screenable normal space is paracompact. Clearly *DC* implies *NC*. Tamano [120] discusses a wide variety of theorems concerning the product invariance of normality and paracompactness and enunciates the following conjecture ***TPC*:** If Y is metrizable and $X \times Y$ is normal then $X \times Y$ is paracompact. Tamano and Morita [83] have shown that to conclude that $X \times Y$ is paracompact it is sufficient to prove $X \times Y$ countably paracompact. Thus Dowker's conjecture implies Tamano's; see Figure 20.

Souslin [113] asked whether a linearly ordered space must be separable whenever it satisfies the countable chain condition (that every disjoint collection of open sets is at most countable). We shall call this conjecture ***SC*;** a counterexample (if it exists) is known as a **Souslin space.** A thorough discussion of this conjecture and related topics is provided by M. E. Rudin [108] who earlier showed [106] that if a Souslin space exists, then so must a counterexample to Dowker's conjecture. In other words, Dowker's conjecture implies Souslin's conjecture. In 1967, Tennenbaum, Solovay, and Jech showed that Souslin's conjecture is consistent with [111] and independent of [54, 122] the axioms of set theory. This result showed also that Dowker's conjecture could not be proved from the present axioms of set theory. Shortly thereafter, M. E. Rudin showed [104] that Dowker's conjecture is actually false.

Epilogue

The concepts and examples discussed in this paper represent not so much the frontier as the established settlements of metrization research. Several papers by Ceder [30], Borges [23], [24], Michael [76], and Worrell and Wicke [138] contain such refinements as M_i-spaces, stratifiable spaces, $\aleph_0$-spaces, and θ bases. In each of these areas there are significant and difficult conjectures similar to those enumerated above; the interested reader can pursue these issues in the papers cited in the bibliography, together with those listed in the excellent bibliographies of [3] and [13]. M. E. Rudin's monograph [107] is an excellent survey of recent work in set-theoretic topology, much of it related to metrization theory. Current research is contained in [99].

Since a metric is a map to the positive reals, it should not be surprising to find that the existence of certain esoteric metrics is intimately related to the existence of certain subsets of the real line. Example 7 provides a very specific instance of this relationship in that potential counterexamples to both Jones' and Dowker's conjectures depend on the existence of certain special subsets of the real line, while the independence theorems of Tall, Silver, Tennenbaum, Solovay, and Jech show that many topological problems depend on fundamentally undecidable problems of set theory. Indeed, recent research in metrization theory has depended heavily on different set-theoretic axioms: the kind of topology you get depends on the kind of models for set theory that you assume.

Two different models are now rather common. In Gödel's constructible universe (where the only sets are those absolutely required by the axioms) Souslin's conjecture is false [55]: there is a nonseparable linearly ordered space that satisfies the countable chain condition. Since the continuum hypothesis holds in the constructible universe, so must Jones' separable Moore space conjecture. Moreover, Fleissner [39] showed that in this universe, every normal space with local basis no greater than c is collectionwise Hausdorff; it follows from this that in the constructible universe every locally compact normal Moore space is metrizable.

The contrasting universe in which much current research takes place is one based on the denial of the continuum hypothesis together with Martin's axiom [70]: no compact Hausdorff space which satisfies the countable chain condition is the union of fewer than c nowhere dense sets. In this universe, Souslin's conjecture is true [111], and every perfectly normal compact space is hereditarily separable [61]. Moreover, there are in Martin's universe many nonmetrizable normal Moore spaces [40], [94], [96], [119]. (In this universe the Cantor tree (Example 14) is normal, even though in the ordinary mathematical universe it is not.) Recently Alster and Przymusiński [7] showed how many of these examples follow from a single property of Martin's universe; they also found, in this universe, a paracompact space whose countable product is normal but not paracompact, and a non-paracompact space whose countable product is perfectly normal.

So despite four decades of intensive work, the normal Moore space conjecture, in the "real" universe of naive set theory, is still unresolved. It is still not known, for instance, whether the normal Moore space conjecture is consistent with the axioms of set theory. Tall (in [117], [119]) showed, using a result of Fleissner, that the existence of metacompact normal nonmetrizable Moore spaces is consistent with CH, and hence with the metrizability of separable normal Moore spaces. Perhaps the nearest thing to a counterexample is the space named George created by Fleissner in 1974 [38] (and simplified by Przymusiński in [95]): George is normal and collectionwise Hausdorff, but not collectionwise normal. A near proof is Reed and Zenor's recent theorem [100] that every locally compact locally connected normal Moore space is metrizable. Yet this theorem may not be very near, after all: it is consistent with the axioms of set theory to assume the existence of nonmetrizable normal Moore spaces that are either locally connected or locally compact, but not both. Research in these areas is now rather intense, so definitive answers may not be far away.

Examples

1. *Open Ordinal Space.* Let X be the set of all ordinal numbers strictly less than the first uncountable ordinal Ω; X carries the interval (or order) topology. Then X is completely collectionwise normal [114], but not fully normal [22].

2. *Closed Ordinal Space.* Let X be the set of all ordinal numbers less than or equal to the first uncountable ordinal Ω. X is compact in the interval topology, but not G_δ since the closed set $\{\Omega\}$ is not a G_δ set. Thus X is neither perfectly normal nor semimetrizable. But of course it is strongly paracompact.

3. *Lower Limit Topology.* Let X be the real line with the topology generated by the sets of the form $[a,b) = \{x \in X \mid a \leqq x < b\}$. Bing [22] cites this space as an example of a regular, separable, strongly screenable (and therefore paracompact) space which is neither perfectly screenable nor developable.

4. *Stratified Plane.* If R is the real line with the Euclidean topology and S is the real line with the discrete topology, then $X = R \times S$ is a nonseparable strongly paracompact metric space.

5. *Bow-Tie Space.* Let X be the Euclidean plane with real axis L. If $d\colon X \times X \to R^+$ is the Euclidean metric on X, we define a semimetric δ

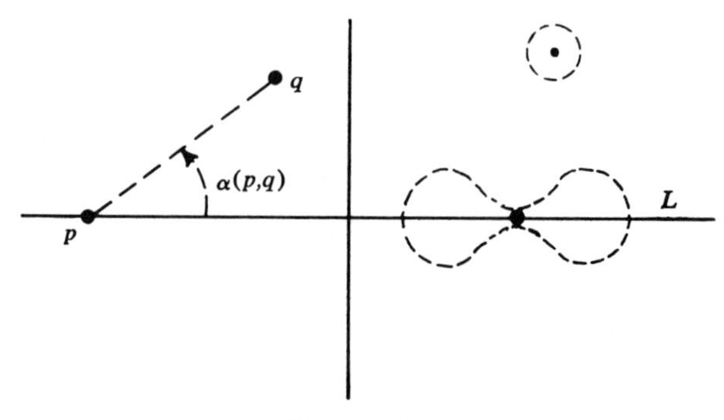

Figure 21.

as follows: $\delta(p,q) = d(p,q)$ if $p,q \in X - L$; $\delta(p,q) = d(p,q) + \alpha(p,q)$ if p or $q \in L$, where $\alpha\,(p,q)$ is the radian measure of the acute angle between L and the line connecting p to q. The topology on X is generated by the semimetric balls of small radius; a neighborhood ball of a point $p \in L$ looks like a bow-tie (Figure 21) or a butterfly, so this space is often called the bow-tie or butterfly space. McAuley [73] introduced this space as an example of a regular semimetric space which is not developable. He showed furthermore that it is paracompact (thus completely collectionwise normal) and hereditarily separable.

6. *Tangent Disc Topology*. Let $P = \{(x,y) \mid x,y \in R,\ y > 0\}$ be the open upper half-plane with the Euclidean topology τ and let L denote the real axis. We generate a topology on $X = P \cup L$ by adding to τ all sets of the form $\{x\} \cup D$, where $x \in L$ and D is an open disc in P which is tangent to L at the point x (Figure 22). This important example was apparently introduced by both Niemytzki (see [13]) and Moore (see [57] as a regular developable space which is not metrizable (since the uncountable closed subset L is discrete and thus not separable in the induced topology). The development which makes X a Moore space is the collection of open balls of radius $1/n$ (including the tangent discs $\{x\} \cup D$ if D has radius

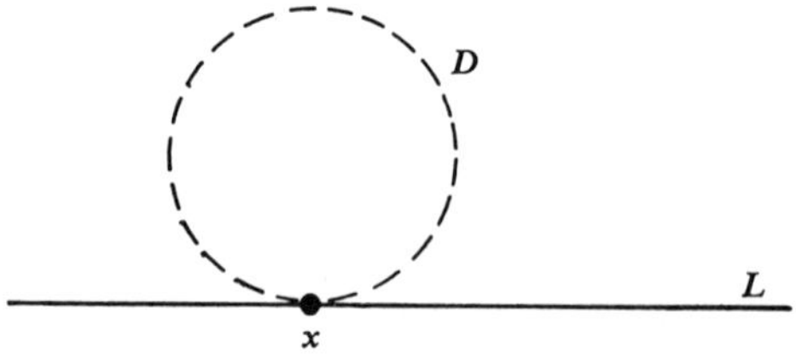

Figure 22.

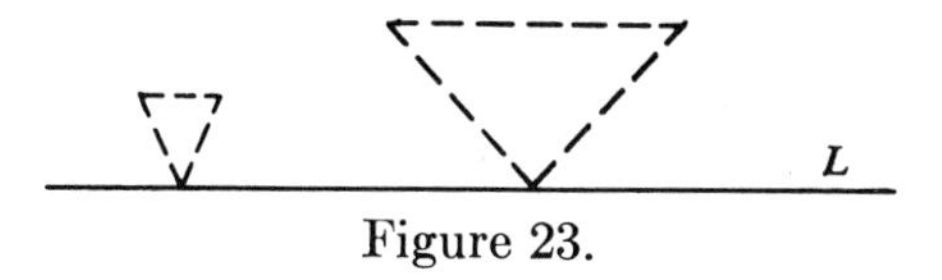

Figure 23.

$1/n$). X is clearly not normal, and neither countably paracompact nor metacompact (see Part II, Example 82).

A common variation (see [58]) of the tangent disc topology is formed by replacing the tangent disc neighborhoods by sets of the form $\{x\} \cup T$ for each $x \in L$, where T is an inverted isosceles triangle in P with vertex at the point x and base parallel to L, such that the radian measure of the vertex angle equals the length of its adjacent sides (Figure 23). McAuley [73] discusses a different variation which is formed from the bow-tie space by rotating each of the bow-tie neighborhoods 90° (Figure 24). Bing [21] introduced a physical model which he called flow space by assuming that water is flowing from left to right across the unit square at the rate of $(1 - x)$ feet per second. Flow space is the closed unit square, and a neighborhood $N_p(t)$ of a point p is the set of all points in X which a swimmer could reach in less than t seconds (Figure 25).

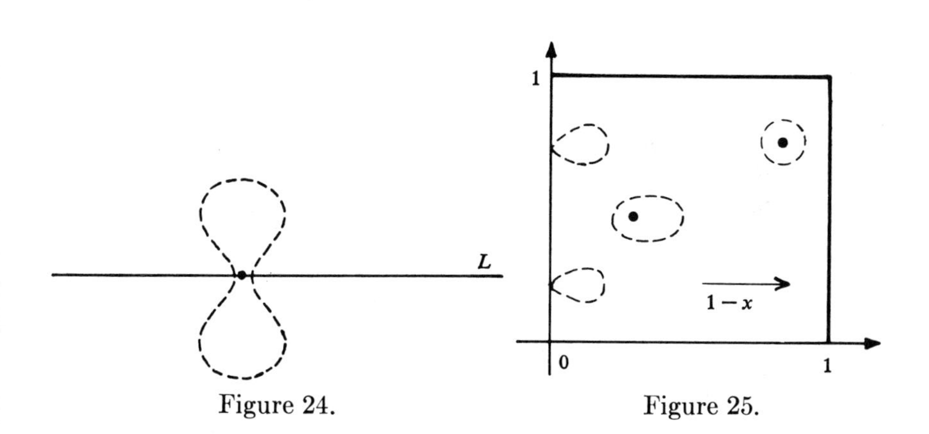

Figure 24. Figure 25.

7. *Tangent Disc Subspaces.* If S is a subset of the real line L, and $Y = P \cup L$ is the tangent disc space, we let X be the subspace $P \cup S$ with the topology induced from Y. The space X is second countable if and only if S is countable, so, since X is regular, X is metrizable if and only if S is countable. X will always be a Moore space since it has the same development as Y, and similarly it will always be separable since the rational lattice points in P are dense in X. Jones [56] showed that every subset of

cardinality c of a separable normal space has a limit point; since S cannot have any limit points, X cannot be normal when S has cardinality c. Bing [22] showed that X is normal if every subset of S is a G_δ-set in (the relative topology of) S; but every uncountable G_δ-subset of the Euclidean real line has cardinality c (by Mazurkiewicz' theorem [68, p. 441]). Thus X would be a normal nonmetrizable Moore space if S were uncountable but of cardinality less than c with the additional property that every subset of S is G_δ in S. Such an S could contain only countable G_δ-subsets of the real line. Clearly the existence of a set with these properties cannot be proved within ordinary set theory since it would constitute a counterexample to the continuum hypothesis. However, Jones [76] constructed a set S of cardinality $\aleph_1$ such that every countable subset of S is G_δ in S.

Younglove [139] studied this example as a possible counterexample to Dowker's conjecture that every countably paracompact space is normal and proved that if S is a G_δ-set, then X is countably paracompact if and only if S is countable. Thus X could be a counterexample to Dowker's conjecture only if S was not a G_δ-subset of the real line L.

8. *Tangent V Topology.* If X is the upper half plane including the real axis L, we let each point of $X - L$ be open and take as a neighborhood basis of points $x \in L$ a "V" with vertex at x, sides of slopes ± 1 and height $1/n$ (Figure 26). Heath [49] showed that X is a metacompact Moore space which is not screenable. Clearly X is neither normal nor separable.

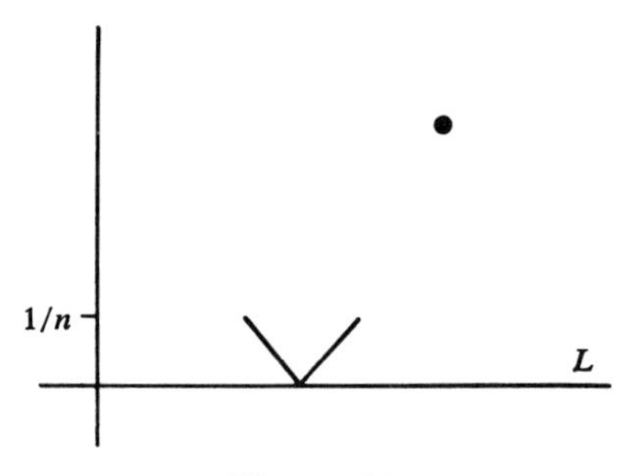

Figure 26.

9. *Picket Fence Topology.* If X is the upper half plane including the real axis L, we let each point of $X - L$ be open, and take as a neighborhood basis of rational points $x \in L$ the vertical line segments of height $1/n$ with lower end point at x. The neighborhood basis of irrational points $x \in L$ consists of line segments of slope 1 and height $1/n$ with their base

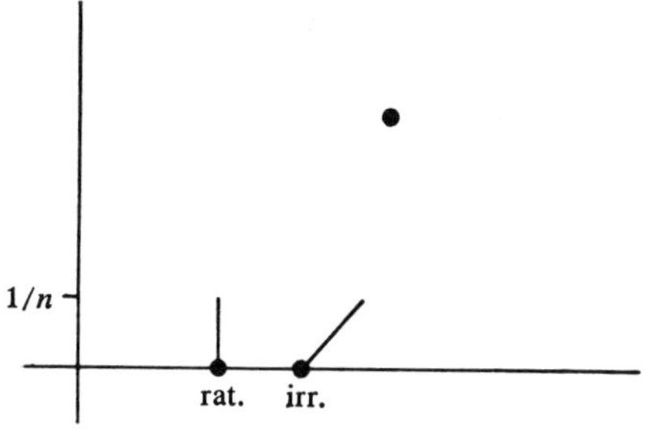

Figure 27.

at the point x (Figure 27). Heath [49] introduced this as a simple example of a screenable Moore space which is not normal.

10. I^I. Let $X = I^I$ be the uncountable Cartesian product of the closed unit interval $I = [0,1]$ with the Tychonoff topology; that is, X is the set of all functions from I to I with the topology of pointwise convergence. Since X is compact and Hausdorff, it is normal; but it is not completely normal (see Part II, Example 105) since it contains a subspace homeomorphic to Z^I, the uncountable product of the positive integers, which Stone [115] showed was not normal. Thus X is strongly paracompact and collectionwise normal but neither perfectly normal nor developable.

11. *Hedgehog.* If K is a cardinal number, a hedgehog X of spininess K is formed from the union of K disjoint copies of the unit interval $[0,1]$ by identifying the zero points of each interval. A metric for X can be defined by $d(x, y) = | x - y |$ if x and y belong to the same segment (or spine), and $d(x,y) = x + y$ otherwise. Alexandroff [3] cites a hedgehog of uncountable spininess as an example of a metric space which is not strongly paracompact.

12. *Bing's Power Space.* If S is some uncountable set with power set P, let $X = \pi_{\lambda \epsilon P}\{0,1\}_\lambda$, where $\{0,1\}_\lambda$ is a copy of the two point discrete space. (If we let 2 denote the two point discrete space, we have $X = 2^{2^S}$.) Since the elements of X are collections of subsets of S, each ultrafilter on S is a point in X; let M denote the subset of X consisting of all principal ultrafilters of S. Then if x_s is the point in X whose λ-th coordinate $(x_s)_\lambda$ equals 1 if and only if $s \in \lambda$, we have $M = \{x_s \in X \mid s \in S\}$. If X has the Tychonoff topology τ, $X - M$ is dense in X. Bing [22] generated a new topology on X by adding to τ all points of $X - M$ as open sets; we shall denote the topology thus generated by σ. M inherits from (X,σ) the discrete topology; furthermore, any two disjoint closed subsets of M are

contained in disjoint open subsets of X (see Part II, Example 142). It follows that X is normal but not perfectly normal [22], metacompact (Part II, Example 142), or collectionwise normal (since it is not even collectionwise Hausdorff: M is an uncountable discrete collection of points without disjoint open neighborhoods for all of its points).

13. *Michael's Power Subspace.* If $X = 2^{2^S}$ is Bing's Power Space, we let Y be the subspace $M \cup L$, where M is the subset of all principal ultrafilters of S and L is the collection of all x in $X - M$ such that $(x)_\lambda = 1$ for only finitely many λ. Michael [77] selected this subspace as an example of normal metacompact space which is not collectionwise normal.

14. *Cantor Tree.* Let C denote the Cantor set in the unit interval [0,1]; the midpoints of the components of $[0,1] - C$ are 1/2, 1/6, 5/6,1/18, 5/18, etc. Let D be the tree (or dendron) in the lower half plane whose vertices are $(1/2,-1)$, $(1/6,-1/2)$, $(5/6,-1/2)$, $(1/18,-1/4)$, $(5/18, -1/4)$, etc. Then the space X is defined as $D \cup C$ (Figure 28), where D inherits the Euclidean topology from the plane, while a basis neighborhood of a point $c \in C$ is a path Γ in the tree D whose upper limit is the point c, together with open segments at each branch point of Γ sufficiently short to avoid including any other branch point. Jones [59] cites this example of Moore as the first example of a nonmetrizable Moore space. The fact that X is nonmetrizable follows from the observation that it is separable but not perfectly separable. Jones [59] shows that X is not normal.

15. *Moore's Road Space.* Let two roads start at the origin of the plane and proceed in opposite directions for one mile each. Let each then branch into two roads which continue for one mile each before each of these now branches into two roads. Continue this in such a way that none of the new roads ever intersect, and so that all roads proceed indefinitely far from the origin. This process generates c roads; at the "end" of each

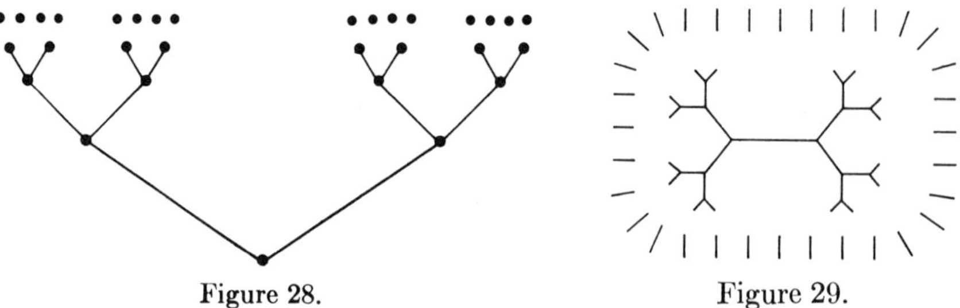

Figure 28. Figure 29.

we adjoin a straight ray of infinite length. This collection of roads is the space X (Figure 29), and we generate a topology from a basis of open discs. This "automobile road" space was introduced by Moore as a graphic variation of the Cantor tree (Example 14); it has the same properties [59].

Figure 30.

1 = *example has property*

0 = *example does not have property*

Example	Regular	Normal	Comp. Normal	Perf. Normal	Coll. Normal	Count. Col. Nor.	Comp. Col. Nor.	Metrizable	Separable	2nd Count.	1st Count.	Hered. Separ.	Lindelöf	Developable	Moore	Semimetric	G_δ	Fully Normal	Str. Paracompact	Paracompact	Metacompact	Count. Paracmpt.	Binormal	Screenable	Str. Screenable	Perf. Screenable	σ-loc. Fin. Base	Uniform Base	Pt. Count. Base
1 Open Ordinal Space	1	1	1	1	1	1	1	0	0	0	1	0	0	0	0	0	1	0	0	0	0	1	1	0	0	0	0	0	
2 Closed Ordinal Space	1	1	1	0	1	1	1	0	0	0	0	0	1	0	0	0	0	1	1	1	1	1	1	0	0	0	0	0	
3 Lower Limit	1	1	1	1	1	1		0	1	0	1		1	0	0	0	1	1	1	1	1	1	1	1	1	0	0	0	
4 Stratified Plane	1	1	1	1	1	1	1	1	0	0	1	0	0	1	1	1	1	1	1	1	1	1	1	1	1	1	1	1	1
5 Bow-Tie	1	1	1	1	1	1	1	0	1	0	1	1	1	0	0	1	1	1	1	1	1	1	1	1	1	0	0	0	
6 Tangent Disc	1	0	0	0	0	0	0	0	1	0	1	0	0	1	1	1	1	0	0	0	0	0	0	0	0	0	0	0	
8 Tangent V	1	0	0	0	0	0	0	0	0	0	1	0	0	1	1	1	1	0	0	0	1		0	0	0	0	0	1	1
9 Picket Fence	1	0	0	0	0	0	0	0	0	0	1	0	0	1	1	1	1	0	0	0	1		0	1	0	0	0	1	1
10 I^I	1	1	0	0	1	1	0	0	1	0	0		1	0	0	0	0	1	1	1	1	1	1	1	1	0	0	0	
11 Hedgehog	1	1	1	1	1	1	1	1	0	0	1	0	0	1	1	1	1	1	0	1	1	1	1	1	1	1	1	1	1
12 Power Space	1	1		0	0		0	0		0			0	0	0	0	0	0	0	0	0				0	0	0	0	
13 Power Subspace	1	1		0	0		0	0		0					0	0	0	0	0	0	1	1	1	0	0	0	0		
14 Cantor Tree	1	0	0	0	0	0	0	0	1	0	1		0	1	1	1	1	0	0	0				0	0	0	0		
15 Moore's Road Space	1	0	0	0	0	0	0	0	0	0	1	0	0	1	1	1	1	0	0	0				0	0	0	0		

PART IV
Appendices

SPECIAL REFERENCE CHARTS

The next few pages contain six basic reference charts which display the properties of the various examples. The properties of a topological space have been grouped into six nearly disjoint categories: separation, compactness, paracompactness, connectedness, disconnectedness, and metrization. In each category we have listed those spaces whose behavior is particularly appropriate. We usually chose any space which represented a counterexample in that category or which exhibited either an unusual or an instructive pathology; occasionally we listed a space simply because it was so well behaved.

Entries in the charts are either 1, 0, or -, meaning, respectively, that the space has the property, does not have the property, or that the property is inapplicable. Occasional blanks represent properties which were not discussed in the text and which do not appear to follow simply from anything that was discussed. Examples are listed by number, and in a few cases the tables extend beyond one page in length.

Table I

SEPARATION AXIOM CHART

	T_0	T_1	T_2	$T_{2\frac{1}{2}}$	T_3	$T_{3\frac{1}{2}}$	T_4	T_5	Semiregular	Regular	Urysohn	Completely Regular	Normal	Completely Normal	Perfectly Normal
4	0	0	0	0	1	1	1	1	0	0	0	0	0	0	0
8	1	0	0	0	0	0	0	0	0	0	0	0	0	0	0
13	1	0	0	0	0	0	1	1	0	0	0	0	0	0	0
17	1	0	0	0	0	0	1	1	0	0	0	0	0	0	0
18	1	1	0	0	0	0	0	0	0	0	0	0	0	0	0
21	0	0	0	0	0	0	0	0	0	0	0	0	0	0	0
24	1	1	1	1	1	1	1	1	1	1	1	1	1	1	0
28	1	1	1	1	1	1	1	1	1	1	1	1	1	1	1
43	1	1	1	1	1	1	1	1	1	1	1	1	1	1	0
52	0	0	0	0	0	0	1	1	0	0	0	0	0	0	0
53	1	0	0	0	0	0	0	0	0	0	0	0	0	0	0
54	0	0	0	0	0	0	0	0	0	0	0	0	0	0	0
55	1	0	0	0	0	0	1	1	0	0	0	0	0	0	0
57	1	0	0	0	0	0	1	0	0	0	0	0	0	0	0
60	1	1	1	0	0	0	0	0	1	0	0	0	0	0	0
63	1	1	1	1	0	0	0	0	0	0	1	0	0	0	0
66	1	1	1	1	0	0	0	0	0	0	1	0	0	0	0
75	1	1	1	0	0	0	0	0	0	0	0	0	0	0	0
78	1	1	1	1	0	0	0	0	0	0	1	0	0	0	0
79	1	1	1	1	0	0	0	0	0	0	1	0	0	0	0
80	1	1	1	1	0	0	0	0	1	0	0	0	0	0	0
81	1	1	1	0	0	0	0	0	0	0	0	0	0	0	0
82	1	1	1	1	1	1	0	0	1	1	1	1	0	0	0
84	1	1	1	1	1	1	0	0	1	1	1	1	0	0	0
86	1	1	1	1	1	1	1	0	1	1	1	1	1	0	0
88	1	1	1	1	0	0	0	0	1	0	1	0	0	0	0
90	1	1	1	1	1	0	0	0	1	1	0	0	0	0	0
91	1	1	1	1	1	0	0	0	1	1	1	0	0	0	0
92	1	1	1	1	1	0	0	0	1	1	0	0	0	0	0
93	1	1	1	1	1	1	0	0	1	1	1	1	0	0	0
94	1	1	1	1	1	0	0	0	1	1	0	0	0	0	0
100	1	1	1	0	0	0	0	0	1	0	0	0	0	0	0
101	1	1	1	1	1	1	1	1	1	1	1	1	1	1	0
103	1	1	1	1	1	1	0	0	1	1	1	1	0	0	0
105	1	1	1	1	1	1	1	0	1	1	1	1	1	0	0
126	1	1	1	1	0	0	0	0	0	0	0	0	0	0	0

Table II

COMPACTNESS CHART

	Compact	σ-Compact	Lindelöf	Countably Compact	Sequentially Compact	Weakly Countably Compact	Pseudocompact	Locally Compact	Strong Local Compactness	σ-Locally Compact	Separable	Second Countable	First Countable	Countable Chain	Metrizable
2	0	1	1	0	0	0	0	1	1	1	1	1	1	1	1
3	0	0	0	0	0	0	0	1	1	0	0	0	1	0	1
6	0	1	1	0	0	1	0	1	1	1	1	1	1	1	0
9	0	1	1	0	0	0	1	1	0	1	1	1	1	1	0
10	0	0	0	0	0	0	1	1	0	0	1	0	1	1	0
14	1	1	1	1	1	1	1	1	1	1	1	1	1	1	0
15	1	1	1	1	1	1	1	1	1	1	0	0	1	0	0
19	1	1	1	1	1	1	1	1	1	1	1	0	0	1	0
20	0	0	1	0	0	0	1	0	0	0	0	0	0	1	0
21	0	0	1	0	0	1	1	0	0	0	0	0	0	1	0
22	1	1	1	1	1	1	1	1	1	1	1	1	1	1	0
25	0	0	1	0	0	0	0	0	0	0	0	0	0	0	0
26	0	1	1	0	0	0	0	0	0	0	1	0	0	1	0
28	0	1	1	0	0	0	0	1	1	1	1	1	1	1	1
30	0	1	1	0	0	0	0	0	0	0	1	1	1	1	1
31	0	0	1	0	0	0	0	0	0	0	1	1	1	1	1
36	0	0	1	0	0	0	0	0	0	0	1	1	1	1	1
42	0	0	0	1	1	1	1	1	1	0	0	0	1	0	0
43	1	1	1	1	1	1	1	1	1	1	0	0	0	0	0
47	0	0	0	1	1	1	1	0	0	0	0	0	0	0	0
48	1	1	1	1	1	1	1	1	1	1	0	0	1	0	0
50	0	1	1	0	0	1	1	1	0	1	1	1	1	1	0
51	0	0	1	0	0	0	0	0	0	0	1	0	1	1	0
52	0	1	1	0	0	1	1	1	0	1	1	1	1	1	0
57	0	1	1	0	0	1	1	1	0	1	1	1	1	1	0
58	0	1	1	0	0	0	0	0	0	0	1	1	1	1	1
60	0	1	1	0	0	0	1	0	0	0	1	1	1	1	0
63	0	0	1	0	0	0	0	0	0	0	0	0	0	1	0
65	0	0	0	0	0	0	0	1	1	0	1	0	1	1	0

Table II (*continued*)

	Compact	σ-Compact	Lindelöf	Countably Compact	Sequentially Compact	Weakly Countably Compact	Pseudocompact	Locally Compact	Strong Local Compactness	σ-Locally Compact	Separable	Second Countable	First Countable	Countable Chain	Metrizable
66	0	0	1	0	0	0	0	0	0	0	1	1	1	1	0
68	0	0	0	0	0	0	0	0	0	0	1	0	1	1	0
70	0	0	1	0	0	0	0	0	0	0	1	1	1	1	1
71	0	0	0	0	0	0	0	0	0	0	0	0	1	0	0
75	0	1	1	0	0	0	1	0	0	0	1	1	1	1	0
76	0	0	0	0	0	0	0	0	0	0	1	0	1	1	0
77	0	0	1	0	0	0	0	0	0	0	1	0	1	1	0
78	0	0	0	0	0	0	0	0	0	0	1	0	1	1	0
79	0	1	1	0	0	0	0	0	0	0	1	1	1	1	0
82	0	0	0	0	0	0	0	0	0	0	1	0	1	1	0
87	0	0	0	0	0	0	1	1	1	0	0	0	0	0	0
89	0	0	0	0	0	0	0	0	0	0	0	0	0	0	0
97	1	1	1	1	1	1	1	1	1	1	0	0	1	0	0
98	0	1	1	0	0	0	0	0	0	0	1	0	0	1	0
99	1	1	1	1	1	1	1	1	1	1	1	0	0	1	0
100	0	1	1	0	0	0	1	0	0	0	1	1	1	1	0
101	1	1	1	1	1	1	1	1	1	1	0	0	0	0	0
102	0	0	1	0	0	0	0	0	0	0	1	1	1	1	1
103	0	0	0	0	0	0	0	0	0	0	–	0	0	–	0
105	1	1	1	1	0	1	1	1	1	1	1	0	0	1	0
106	0	0	0	1	0	1	1	1	1	0	0	0	0	0	0
107	1	1	1	1	1	1	1	1	1	1	1	0	1	1	0
109	0	0	0	0	0	0	0	0	0	0	0	0	0	0	0
111	1	1	1	1	0	1	1	1	1	1	1	0	0	1	0
112	0	0	0	1		1	1			0	1	0		1	0
113	0	0	0	0	0	0	1	0	0	0	1	0	0	1	0
132	0	0	0	0	0	0	0	0	0	0	0	0	1	0	1
136	0	0	1	0	0	0	0	0	0	0	1	1	1	1	1
139	0	0	0	0	0	0	0	0	0	0	0	0	1	0	1
140	0	0	0	0	0	0	0	0	0	0	0	0	1	0	1
141	0	0	0	0	0	0	0	0	0	0	0	0	0	0	0

Table III

PARACOMPACTNESS CHART

	Compact	Paracompact	Metacompact	Countably Compact	Countably Paracompact	Countably Metacompact	Fully Normal	Fully T_4	Metrizable
28	0	1	1	0	1	1	1	1	1
42	0	0	0	1	1	1	0	0	0
51	0	1	1	0	1	1	1	1	0
54	0	0	1	0	0	1	0	0	0
62	0	1	1	0	1	1	0	1	0
64	0	0	1	0	0	1	0	0	0
78	0	0	0	0	0	1	0	0	0
82	0	0	0	0	0	1	0	0	0
84	0	0	0	0	0	1	0	0	0
85	0	0	1	0		1	0	0	0
89	0	0	1	0	0	1	0	0	0
96	0	0	1	0	0	1	0	0	0
103	0	0		0			0	0	0
143	0	0	1	0	1	1	0	0	0

Table IV

CONNECTEDNESS CHART

	Connected	Path Connected	Arc Connected	Hyperconnected	Ultraconnected	Locally Connected	Locally Path Connected	Locally Arc Connected	Biconnected	Dispersion Point
3	0	0	0	0	0	1	1	1	0	0
9	1	1	0	1	0	1	1	0	1	1
14	1	1	0	0	1	1	1	0	1	1
17	0	0	0	0	0	1	1	0	0	0
18	1	0	0	1	0	1	0	0	0	0
19	1	1	1	1	0	1	1	1	0	0
28	1	1	1	0	0	1	1	1	0	0
35	1	0	0	0	0	0	0	0	1	1
46	1	0	0	0	0	1	0	0	0	0
48	1	0	0	0	0	1	0	0	0	0
52	1	1	1	1	1	1	1	1	0	0
53	1	1	1	1	0	1	1	1	0	0
55	0	0	0	0	0	1	1	0	0	0
56	1	1	0	1	0	1	1	0	0	0
57	1	1	0	0	1	1	1	0	0	0
60	1	0	0	0	0	0	0	0	0	0
61	1	0	0	0	0	1	0	0	0	0
66	1	0	0	0	0	0	0	0	0	0
75	1	0	0	0	0	0	0	0	0	0
101	1	1	1	0	0	0	0	0	0	0
116	1	0	0	0	0	0	0	0	0	0
118	1	1	1	0	0	0	0	0	0	0
119	1	0	0	0	0	0	0	0	0	0
120	1	1	1	0	0	0	0	0	0	0
121	1	1	0	0	1	0	0	0	0	0
126	1	0	0	0	0	0	0	0	1	1
128	1	0	0	0	0	0	0	0	1	1
132	1	1	1	0	0	1	1	1	0	0

Table V

DISCONNECTEDNESS CHART

	T_0	T_1	T_2	Urysohn	Regular	Totally Path Disconnected	Totally Disconnected	Totally Separated	Extremally Disconnected	Zero Dimensional	Scattered
6	0	0	0	0	0	0	0	0	0	1	0
8	1	0	0	0	0	0	0	0	0	0	1
23	1	1	1	1	1	1	1	1	0	1	1
26	1	1	1	1	1	1	1	1	0	1	1
27	1	1	0	0	0	1	1	0	0	0	1
29	1	1	1	1	1	1	1	1	0	1	0
30	1	1	1	1	1	1	1	1	0	1	0
35	1	1	0	0	0	1	0	0	0	0	0
42	1	1	1	1	1	1	1	1	0	1	1
51	1	1	1	1	1	1	1	1	0	1	0
57	1	0	0	0	0	0	0	0	0	0	1
58	1	1	1	1	1	1	1	1	0	1	0
60	1	1	1	0	0	1	0	0	0	0	0
65	1	1	1	1	1	1	1	1	0	1	1
66	1	1	1	1	0	1	0	0	0	0	0
70	1	1	1	1	1	1	1	1	0	1	0
72	1	1	1	1	0	1	1	0	0	0	1
79	1	1	1	1	0	1	1	1	0	0	1
80	1	1	1	0	0	1	1	0	0	0	0
90	1	1	1	0	1	1	1	0	0	0	1
91	1	1	1	1	1	1	1	1	0	0	1
95	1	1	1	1	1	1	1	1	0	1	0
96	1	1	1	1	0	1	1	1	0	0	0
99	1	1	0	0	0	1	1	0	0	0	1
102	1	1	1	1	1	1	1	1	0	1	0
111	1	1	1	1	1	1	1	1	1	1	0
127	1	1	1	1	0	1	1	1	0	0	0
129	1	1	1	1	1	1	1	0	0	0	0
136	1	1	1	1	1	1	1	1	0		0
139	1	1	1	1	1	1	1	1	0	1	1

Table VI

METRIZABILITY CHART

	Metrizable	Regular	Second Countable	σ-Locally Finite Base	Topologically Complete	Second Category	Hausdorff	Compact	Locally Compact	Countably Compact	Paracompact
23	1	1	1	1	1	1	1	1	1	1	1
24	0	1	0	0	–	1	1	1	1	1	1
28	1	1	1	1	1	1	1	0	1	0	1
30	1	1	1	1	0	0	1	0	0	0	1
31	1	1	1	1	1	1	1	0	0	0	1
35	0	0	1	1	–	0	0	1	1	1	1
43	0	1	0	0	–	1	1	1	1	1	1
45	0	1	0	0	–	1	1	0	1	1	0
51	0	1	0	0	–	1	1	0	0	0	1
58	1	1	1	1	0	0	1	0	0	0	1
59	1	1	1	1	0	0	1	0	0	0	1
65	0	1	0	0	–	1	1	0	1	0	0
70	1	1	1	1		1	1	0	0	0	1
71	0	1	0	0	–	1	1	0	0	0	1
75	0	0	1	1	–	0	1	0	0	0	0
83	1	1	1	1		1	1	0	0	0	1
85	0	1	0	0	–	1	1	0	0	0	0
102	1	1	1	1	1	1	1	0	0	0	1
103	0	1	0	0	–		1	0	0	0	0
104	1	1	0	1	1	1	1	0	0	0	1
107	0	1	0	0	–	1	1	1	1	1	1
108	1	1	1	1	1	1	1	0	0	0	1
111	0	1	0	0	–	1	1	1	1	1	1
132	1	1	0	1	1	1	1	0	0	0	1
136	1	1	1	1			1	0	0	0	1
139	1	1	0	1	1	1	1	0	0	0	1
140	1	1	0	1	1	1	1	0	0	0	1
141	0	1	0	0	–	1	1	0	0	0	0

	T_0	T_1	T_2	$T_{2\frac{1}{2}}$	T_3	$T_{3\frac{1}{2}}$	T_4	T_5	Urysohn	Semiregular	Regular	Completely Regular	Normal	Completely Normal	Perfectly Normal	Compact	σ-Compact	Lindelöf	Countably Compact	Sequentially Compact	Weak. Count. Compact	Pseudocompact	Locally Compact	Strong Loc. Compact	σ-Locally Compact	Separable	Second Countable	First Countable	Count. Chain Cond.	Paracompact
1	1	1	1	1	1	1	1	1	1	1	1	1	1	1	1	1	1	1	1	1	1	1	1	1	1	1	1	1	1	1
2	1	1	1	1	1	1	1	1	1	1	1	1	1	1	1	0	1	1	0	0	0	0	1	1	1	1	1	1	1	1
3	1	1	1	1	1	1	1	1	1	1	1	1	1	1	1	0	0	0	0	0	0	0	1	1	0	0	0	1	0	1
4	0	0	0	0	1	1	1	1	0	0	0	0	0	0	0	1	1	1	1	1	1	1	1	1	1	1	1	1	1	1
5																														
6	0	0	0	0	1	1	1	1	0	0	0	0	0	0	0	0	1	1	0	0	1	0	1	1	1	1	1	1	1	1
7																														
8	1	0	0	0	0	0	0	0	0	0	0	0	0	0	0	1	1	1	1	1	1	1	1	1	1	1	1	1	1	1
9	1	0	0	0	0	0	0	0	0	0	0	0	0	0	0	0	1	1	0	0	0	1	1	0	1	1	1	1	1	0
10	1	0	0	0	0	0	0	0	0	0	0	0	0	0	0	0	0	0	0	0	0	1	1	0	0	1	0	1	1	0
11	1	0	0	0	0	0	1	1	0	0	0	0	0	0	0	1	1	1	1	1	1	1	1	1	1	1	1	1	1	1
12																														
13	1	0	0	0	0	0	1	1	0	0	0	0	0	0	0	1	1	1	1	1	1	1	1	1	1	1	1	1	1	1
14	1	0	0	0	0	0	1	1	0	0	0	0	0	0	0	1	1	1	1	1	1	1	1	1	1	1	1	1	1	1
15	1	0	0	0	0	0	1	1	0	0	0	0	0	0	0	1	1	1	1	1	1	1	1	1	1	0	0	1	0	1
16																														
17	1	0	0	0	0	0	1	1	0	0	0	0	0	0	0	1	1	1	1	1	1	1	1	1	1	0	0	1	0	1
18	1	1	0	0	0	0	0	0	0	0	0	0	0	0	0	1	1	1	1	1	1	1	1	1	1	1	1	1	1	1
19	1	1	0	0	0	0	0	0	0	0	0	0	0	0	0	1	1	1	1	1	1	1	1	1	1	1	0	0	1	1
20	1	1	0	0	0	0	0	0	0	0	0	0	0	0	0	0	0	1	0	0	0	1	0	0	0	0	0	0	1	0
21	0	0	0	0	0	0	0	0	0	0	0	0	0	0	0	0	0	1	0	0	1	1	0	0	0	0	0	0	1	0
22	1	1	0	0	0	0	0	0	0	0	0	0	0	0	0	1	1	1	1	1	1	1	1	1	1	1	1	1	1	1

GENERAL REFERENCE CHART

Metacompact	Count. Paracompact	Count. Metacompact	Fully Normal	Fully T_4	Connected	Path Connected	Arc Connected	Hyperconnected	Ultraconnected	Locally Connected	Local Path Connected	Local Arc Connected	Biconnected	Has Dispersion Point	Totally Path Disc.	Totally Disconnected	Totally Separated	Extremally Disc.	Zero Dimensional	Scattered	Discrete	Metrizable	σ-Locally Finite Base	Topol. Complete	Second Category	Countable	Card Less Than c	Card = c	Card Not Exc. 2^c	Strongly Connected
1	1	1	1	1	0	0	0	0	0	1	1	1	0	0	1	1	1	1	1	1	1	1	1	1	1	1	1	0	1	0
1	1	1	1	1	0	0	0	0	0	1	1	1	0	0	1	1	1	1	1	1	1	1	1	1	1	1	1	0	1	0
1	1	1	1	1	0	0	0	0	0	1	1	1	0	0	1	1	1	1	1	1	1	1	1	1	1	0	–	–	–	0
1	1	1	0	1	1	1	–	1	1	1	1	–	–	–	0	0	0	0	1	0	0	0	1	–	1	–	–	–	–	1
1	1	1	0	1	0	0	0	0	0	1	1	0	0	0	0	0	0	0	1	0	0	0	1	–	1	1	1	0	1	0
1	1	1	0	0	1	1	0	1	0	1	1	0	1	1	0	0	0	0	0	1	0	0	1	–	1	1	1	0	1	1
0	0	0	0	0	1	1	0	1	0	1	1	0	1	1	0	0	0	0	0	1	0	0	1	–	1	1	1	0	1	1
0	0	0	0	0	1	1	0	1	0	1	1	0	1	1	0	0	0	0	0	1	0	0	0	–	1	0	–	–	–	1
1	1	1	0	0	1	1	0	1	1	1	1	0	1	1	0	0	0	0	0	1	0	0	1	–	1	1	1	0	1	1
1	1	1	0	1	1	1	0	0	1	1	1	0	1	1	0	0	0	0	0	1	0	0	1	–	1	1	1	0	1	1
1	1	1	0	1	1	1	0	0	1	1	1	0	1	1	0	0	0	0	0	1	0	0	1	–	1	1	1	0	1	1
1	1	1	0	1	1	1	0	0	1	1	1	0	1	1	0	0	0	0	0	1	0	0	1	–	1	0	–	–	–	1
1	1	1	0	1	0	0	0	0	0	1	1	0	0	0	0	0	0	0	0	1	0	0	1	–	1	0	0	1	1	0
1	1	1	0	0	1	0	0	1	0	1	0	0	0	0	1	0	0	0	0	0	0	0	1	–	0	1	1	0	1	1
1	1	1	0	0	1	1	1	1	0	1	1	1	0	0	0	0	0	0	0	0	0	0	0	–	1	0	–	–	–	1
0	0	0	0	0	1	0	0	1	0	1	0	0	0	0	1	0	0	0	0	0	0	0	0	–	1	0	–	–	–	1
0	0	0	0	0	1	0	0	1	0	1	0	0	0	0	0	0	0	0	0	0	0	0	0	–	1	0	–	–	–	1
1	1	1	0	0	1	1	1	1	0	1	1	1	0	0	0	0	0	0	0	0	0	0	1	–	0	0	0	1	1	1

	T_0	T_1	T_2	$T_{2\frac{1}{2}}$	T_3	$T_{3\frac{1}{2}}$	T_4	T_5	Urysohn	Semiregular	Regular	Completely Regular	Normal	Completely Normal	Perfectly Normal	Compact	σ-Compact	Lindelöf	Countably Compact	Sequentially Compact	Weak. Count. Compact	Pseudocompact	Locally Compact	Strong Loc. Compact	σ-Locally Compact	Separable	Second Countable	First Countable	Count. Chain Cond.	Paracompact
23	1	1	1	1	1	1	1	1	1	1	1	1	1	1	1	1	1	1	1	1	1	1	1	1	1	1	1	1	1	1
24	1	1	1	1	1	1	1	1	1	1	1	1	1	1	0	1	1	1	1	1	1	1	1	1	1	0	0	0	1	1
25	1	1	1	1	1	1	1	1	1	1	1	1	1	1	0	0	0	1	0	0	0	0	0	0	0	0	0	0	0	1
26	1	1	1	1	1	1	1	1	1	1	1	1	1	1	1	0	1	1	0	0	0	0	0	0	0	1	0	0	1	1
27	1	1	0	0	0	0	0	0	0	0	0	0	0	0	0	1	1	1	1	1	1	1	1	1	1	0	0	0	0	1
28	1	1	1	1	1	1	1	1	1	1	1	1	1	1	1	0	1	1	0	0	0	0	1	1	1	1	1	1	1	1
29	1	1	1	1	1	1	1	1	1	1	1	1	1	1	1	1	1	1	1	1	1	1	1	1	1	1	1	1	1	1
30	1	1	1	1	1	1	1	1	1	1	1	1	1	1	1	0	1	1	0	0	0	0	0	0	0	1	1	1	1	1
31	1	1	1	1	1	1	1	1	1	1	1	1	1	1	1	0	0	1	0	0	0	0	0	0	0	1	1	1	1	1
32																														
33																														
34																														
35	1	1	0	0	0	0	0	0	0	0	0	0	0	0	0	1	1	1	1	1	1	1	1	1	1	1	0	0	1	1
36	1	1	1	1	1	1	1	1	1	1	1	1	1	1	1	0	0	1	0	0	0	0	0	0	0	1	1	1	1	1
37																														
38	1	1	1	1	1	1	1	1	1	1	1	1	1	1	1	1	1	1	1	1	1	1	1	1	1	1	1	1	1	1
39																														
40	1	1	1	1	1	1	1	1	1	1	1	1	1	1	1	0	1	1	0	0	0	0	1	1	1	1	1	1	1	1
41	1	1	1	1	1	1	1	1	1	1	1	1	1	1	1	1	1	1	1	1	1	1	1	1	1	1	1	1	1	1
42	1	1	1	1	1	1	1	1	1	1	1	1	1	1	0	0	0	0	1	1	1	1	1	1	0	0	0	1	0	0
43	1	1	1	1	1	1	1	1	1	1	1	1	1	1	0	1	1	1	1	1	1	1	1	1	1	0	0	0	0	1
44																														
45	1	1	1	1	1	1	1	1	1	1	1	1	1	1	0	0	0	0	1	1	1	1	1	1	0	0	0	1	0	0
46	1	1	1	1	1	1	1	1	1	1	1	1	1	1	0	1	1	1	1	1	1	1	1	1	1	0	0	0	0	1
47	1	1	1	1	0	0	0	0	1	0	0	0	0	0	0	0	0	0	1	1	1	1	0	0	0	0	0	0	0	0
48	1	1	1	1	1	1	1	1	1	1	1	1	1	1	0	1	1	1	1	1	1	1	1	1	1	0	0	1	0	1
49																														
50	1	0	0	0	0	0	1	1	0	0	0	0	0	0	0	0	1	1	0	0	1	1	1	0	1	1	1	1	1	0
51	1	1	1	1	1	1	1	1	1	1	1	1	1	1	1	0	0	1	0	0	0	0	0	0	0	1	0	1	1	1
52	0	0	0	0	0	0	1	1	0	0	0	0	0	0	0	0	1	1	0	0	1	1	1	0	1	1	1	1	1	0
53	1	0	0	0	0	0	0	0	0	0	0	0	0	0	0	1	1	1	1	1	1	1	1	1	1	1	1	1	1	1
54	0	0	0	0	0	0	0	0	0	0	0	0	0	0	0	0	1	1	0	0	0	1	1	0	1	1	1	1	1	0
55	1	0	0	0	0	0	1	1	0	0	0	0	0	0	0	0	1	1	0	0	0	0	1	1	1	1	1	1	1	1
56	1	0	0	0	0	0	0	0	0	0	0	0	0	0	0	1	1	1	1	1	1	1	1	1	1	1	1	1	1	1
57	1	0	0	0	0	0	1	0	0	0	0	0	0	0	0	0	1	1	0	0	1	1	1	0	1	1	1	1	1	0
58	1	1	1	1	1	1	1	1	1	1	1	1	1	1	1	0	1	1	0	0	0	0	0	0	0	1	1	1	1	1

GENERAL REFERENCE CHART (continued)

Metacompact	Count. Paracompact	Count. Metacompact	Fully Normal	Fully T_4	Connected	Path Connected	Arc Connected	Hyperconnected	Ultraconnected	Locally Connected	Local Path Connected	Local Arc Connected	Biconnected	Has Dispersion Point	Totally Path Disc.	Totally Disconnected	Totally Separated	Extremally Disc.	Zero Dimensional	Scattered	Discrete	Metrizable	σ-Locally Finite Base	Topol. Complete	Second Category	Countable	Card Less Than c	Card = c	Card Not Exc. 2^c	Strongly Connected
1	1	1	1	1	0	0	0	0	0	0	0	0	0	0	1	1	1	0	1	1	0	1	1	1	1	1	1	0	1	0
1	1	1	1	1	0	0	0	0	0	0	0	0	0	0	1	1	1	0	1	1	0	0	0	–	1	0	–	–	–	0
1	1	1	1	1	0	0	0	0	0	0	0	0	0	0	1	1	1	0	1	1	0	0	0	–	1	0	–	–	–	0
1	1	1	1	1	0	0	0	0	0	0	0	0	0	0	1	1	1	0	1	1	0	0	0	–	1	1	1	0	1	0
1	1	1	0	0	0	0	0	0	0	0	0	0	0	0	1	1	0	0	0	1	0	0	0	–	1	0	–	–	–	0
1	1	1	1	1	1	1	1	0	0	1	1	1	0	0	0	0	0	0	0	0	0	1	1	1	1	0	0	1	1	0
1	1	1	1	1	0	0	0	0	0	0	0	0	0	0	1	1	1	0	1	0	0	1	1	1	1	0	0	1	1	0
1	1	1	1	1	0	0	0	0	0	0	0	0	0	0	1	1	1	0	1	0	0	1	1	0	0	1	1	0	1	0
1	1	1	1	1	0	0	0	0	0	0	0	0	0	0	1	1	1	0	1	0	0	1	1	1	1	0	0	1	1	0
1	1	1	0	0	1	0	0	0	0	0	0	0	1	1	1	0	0	0	0	0	0	0	1	–	0	1	1	0	1	1
1	1	1	1	1	1	1	1	0	0	1	1	1	0	0	0	0	0	0	0	0	0	1	1	1	1	0	0	1	1	0
1	1	1	1	1	1	1	1	0	0	1	1	1	0	0	0	0	0	0	0	0	0	1	1	1	1	0	0	1	1	0
1	1	1	1	1	0	0	0	0	0	0	0	0	0	0	1	1	1	0	1	1	0	1	1	1	1	1	1	0	1	0
1	1	1	1	1	0	0	0	0	0	0	0	0	0	0	1	1	1	0	1	1	0	1	1	1	1	1	1	0	1	0
0	1	1	0	0	0	0	0	0	0	0	0	0	0	0	1	1	1	0	1	1	0	0	0	–	1	0	–	–	1	0
1	1	1	1	1	0	0	0	0	0	0	0	0	0	0	1	1	1	0	1	1	0	0	0	–	1	0	–	–	1	0
0	1	1	0	0	1	1	1	0	0	1	1	1	0	0	0	0	0	0	0	0	0	0	0	–	1	0	0	1	1	0
1	1	1	1	1	1	0	0	0	0	1	0	0	0	0	0	0	0	0	0	0	0	0	0	–	1	0	0	1	1	0
0	1	1	0	0	1	0	0	0	0	0	0	0	0	0	0	0	0	0	0	0	0	0	0	–	1	0	0	1	1	0
1	1	1	1	1	1	0	0	0	0	1	0	0	0	0	0	0	0	0	0	0	0	0	0	–	1	0	0	1	1	0
0	0	0	0	0	1	1	1	1	1	1	1	1	0	0	0	0	0	0	0	0	0	0	1	–	0	0	0	1	1	1
1	1	1	1	1	0	0	0	0	0	0	0	0	0	0	1	1	1	0	1	0	0	0	0	–	1	0	0	1	1	0
0	0	0	0	0	1	1	1	1	1	1	1	1	0	0	0	0	0	0	0	0	0	0	1	–	0	0	0	1	1	1
1	1	1	0	0	1	1	1	1	0	1	1	1	0	0	0	0	0	0	0	0	0	0	1	–	1	0	0	1	1	1
1	0	1	0	0	1			1	0	1			0	0	0	0	0	0	0	0	0	0	1	–	0	0	0	1	1	1
1	1	1	0	1	0	0	0	0	0	1	1	0	0	0	0	0	0	0	0	1	0	0	1	–	1	1	1	0	1	0
1	1	1	0	0	1	1	0	1	0	1	1	0	0	0	0	0	0	0	0	0	0	0	1	–	1	1	1	0	1	1
0	0	0	0	0	1	1	0	0	1	1	1	0	0	0	0	0	0	0	0	1	0	0	1	–	1	1	1	0	1	1
1	1	1	1	1	0	0	0	0	0	0	0	0	0	0	1	1	1	0	1	0	0	1	1	0	0	1	1	0	1	0

	T_0	T_1	T_2	$T_{2\frac{1}{2}}$	T_3	$T_{3\frac{1}{2}}$	T_4	T_5	Urysohn	Semiregular	Regular	Completely Regular	Normal	Completely Normal	Perfectly Normal	Compact	σ-Compact	Lindelöf	Countably Compact	Sequentially Compact	Weak. Count. Compact	Pseudocompact	Locally Compact	Strong Loc. Compact	σ-Locally Compact	Separable	Second Countable	First Countable	Count. Chain Cond.	Paracompact
59	1	1	1	1	1	1	1	1	1	1	1	1	1	1	1	0	1	1	0	0	0	0	0	0	0	1	1	1	1	1
60	1	1	1	0	0	0	0	0	0	1	0	0	0	0	0	0	1	1	0	0	0	1	0	0	0	1	1	1	1	0
61	1	1	1	0	0	0	0	0	0		0	0	0	0	0	0	1	1	0	0	0	1	0	0	0	1	1	1	1	0
62	0	0	0	0	1	1	1	1	0	0	0	0	0	0	0	0	1	1	0	0	1	0	1	1	1	1	1	1	1	1
63	1	1	1	1	0	0	0	0	1	0	0	0	0	0	0	0	0	1	0	0	0	0	0	0	0	0	0	0	1	0
64	1	1	1	1	0	0	0	0	1	0	0	0	0	0	0	0	1	1	0	0	0	0	0	0	0	1	1	1	1	0
65	1	1	1	1	1	1	0	0	1	1	1	1	0	0	0	0	0	0	0	0	0	0	1	1	0	1	0	1	1	0
66	1	1	1	1	0	0	0	0	1	0	0	0	0	0	0	0	0	1	0	0	0	0	0	0	0	1	1	1	1	0
67	1	1	1	1	0	0	0	0	1	0	0	0	0	0	0	0	0	1	0	0	0	0	0	0	0	1	1	1	1	0
68	1	1	1	1	0	0	0	0	1	0	0	0	0	0	0	0	0	0	0	0	0	0	0	0	0	1	0	1	1	0
69	1	1	1	1	0	0	0	0	1	0	0	0	0	0	0	0	0	1	0	0	0	0	0	0	0	1	1	1	1	0
70	1	1	1	1	1	1	1	1	1	1	1	1	1	1	1	0	0	1	0	0	0	0	0	0	0	1	1	1	1	1
71	1	1	1	1	1	1	1	1	1	1	1	1	1	1	0	0	0	0	0	0	0	0	0	0	0	0	0	1	0	1
72	1	1	1	1	0	0	0	0	1		0	0	0	0	0	0	0	0	0	0	0	0	0	0	0	1	0	1	1	0
73	1	1	0	0	0	0	0	0	0	0	0	0	0	0	0	1	1	1	1	1	1	1	1	1	1	1	1	1	1	1
74	1	1	1	0	0	0	0	0	0	1	0	0	0	0	0	0	1	1	0	0	0	0	0	0	0	1	1	1	1	0
75	1	1	1	0	0	0	0	0	0	0	0	0	0	0	0	0	1	1	0	0	0	1	0	0	0	1	1	1	1	0
76	1	1	1	1	0	0	0	0	1	0	0	0	0	0	0	0	0	0	0	0	0	0	0	0	0	1	0	1	1	0
77	1	1	1	1	0	0	0	0	1	0	0	0	0	0	0	0	0	1	0	0	0	0	0	0	0	1	0	1	1	0
78	1	1	1	1	0	0	0	0	1	0	0	0	0	0	0	0	0	0	0	0	0	0	0	0	0	1	0	1	1	0
79	1	1	1	1	0	0	0	0	1	0	0	0	0	0	0	0	1	1	0	0	0	0	0	0	0	1	1	1	1	0
80	1	1	1	1	0	0	0	0	0	1	0	0	0	0	0	0	1	1	0	0	0	0	0	0	0	1	1	1	1	0
81	1	1	1	0	0	0	0	0	0	0	0	0	0	0	0	0	1	1	0	0	0	0	0	0	0	1	1	1	1	0
82	1	1	1	1	1	1	0	0	1	1	1	1	0	0	0	0	0	0	0	0	0	0	0	0	0	1	0	1	1	0
83	1	1	1	1	1	1	1	1	1	1	1	1	1	1	1	0		1	0	0	0	0	0	0	0	1	1	1	1	1
84	1	1	1	1	1	1	0	0	1	1	1	1	0	0	0	0	0	0	0	0	0	0	0	0	0	1	0	1	1	0
85	1	1	1	1	1	1	0	0	1	1	1	1	0	0	0	0	0	0	0	0	0	0	0	0	0	0	0	1	0	0
86	1	1	1	1	1	1	1	0	1	1	1	1	1	0	0	1	1	1	1		1	1	1	1	1	0	0	0	0	1
87	1	1	1	1	1	1	0	0	1	1	1	1	0	0	0	0	0	0	0	0	0	1	1	1	0	0	0	0	0	0
88	1	1	1	1	0	0	0	0	1	1	0	0	0	0	0	0	0	0	0	0	0		0	0	0	0	0	0	0	0
89	1	1	1	1	1	1	0	0	1	1	1	1	0	0	0	0	0	0	0	0	0	0	0	0	0	0	0	0	0	0
90	1	1	1	1	1	0	0	0	0	1	1	0	0	0	0	0	0	0	1	1	1	1	0	0	0	0	0	0	0	0
91	1	1	1	1	1	0	0	0	1	1	1	0	0	0	0	0	0	0	0	0	0		0	0	0	0	0	0	0	0
92	1	1	1	1	1	0	0	0	0	1	1	0	0	0	0	0	0	0				1	0	0	0		0			0
93	1	1	1	1	1	1	0	0	1	1	1	1	0	0	0	0	0	0	0	0	0	0	1	1	0	0	0	0	0	0
94	1	1	1	1	1	0	0	0	0	1	1	0	0	0	0	0	0	0	0	0	0	0	0	0	0	0	0	0	0	0

GENERAL REFERENCE CHART (continued)

Metacompact	Count. Paracompact	Count. Metacompact	Fully Normal	Fully T_4	Connected	Path Connected	Arc Connected	Hyperconnected	Ultraconnected	Locally Connected	Local Path Connected	Local Arc Connected	Biconnected	Has Dispersion Point	Totally Path Disc.	Totally Disconnected	Totally Separated	Extremally Disc.	Zero Dimensional	Scattered	Discrete	Metrizable	σ-Locally Finite Base	Topol. Complete	Second Category	Countable	Card Less Than c	Card = c	Card Not Exc. 2^c	Strongly Connected
1	1	1	1	1	0	0	0	0	0	0	0	0	0	0	1	1	1	0	1	0	0	1	1	0	0	1	1	0	1	0
1	0	1	0	0	1	0	0	0	0	0	0	0	0	0	1	0	0	0	0	0	0	0	1	–		1	1	0	1	1
1	0	1	0	0	1	0	0	0	0	1	0	0	0	0	1	0	0	0	0	0	0	0	1	–		1	1	0	1	1
1	1	1	0	1	1	1	1	0	0	1	1	1	0	0	0	0	0	0	0	0	0	0	1	–	0	0	0	1	1	0
0	0	0	0	0	1	0	0	0	0	1	0	0	0	0	1	0	0	0	0	0	0	0		–		0	0	1	1	0
1	0	1	0	0	1	0	0	0	0	0	0	0	0	0	0	0	0	0	0	0	0	0	1	–	1	0	0	1	1	0
0		1	0	0	0	0	0	0	0	0	0	0	0	0	1	1	1	0	1	1	0	0	0	–	1	0	0	1	1	0
1	0	1	0	0	1	0	0	0	0	0	0	0	0	0	1	0	0	0	0	0	0	0	1	–		0	0	1	1	0
1	0	1	0	0	1	0	0	0	0	0	0	0	0	0	1	0	0	0	0	0	0	0	1	–		0	0	1	1	0
0	0	0	0	0	1	0	0	0	0	0	0	0	0	0	1	0	0	0	0	0	0	0		–		0	0	1	1	0
0	0	0	0	0	1	0	0	0	0	0	0	0	0	0	1	0	0	0	0	0	0	0	1	–		0	0	1	1	0
1	1	1	1	1	0	0	0	0	0	0	0	0	0	0	1	1	1	0	1	0	0	1	1		1	0	0	1	1	0
1	1	1	1	1	0	0	0	0	0	0	0	0	0	0	1	1	1	0	1	0	0	0	0	–	1	0	0	1	1	0
0	0	1	0	0	0	0	0	0	0	0	0	0	0	0	1	1	0	0	0	1	0	0		–		0	0	1	1	0
1	1	1	0	0	1	1	1	0	0	1	1	1	0	0	0	0	0	0	0	0	0	0	1	–	1	0	0	1	1	0
1	0	1	0	0	1	1	1	0	0	1	1	1	0	0	0	0	0	0	0	0	0	0	1	–	1	0	0	1	1	0
1	0	1	0	0	1	0	0	0	0	0	0	0	0	0	1	0	0	0	0	0	0	0	1	–	0	1	1	0	1	1
0	0	1	0	0	1	1	1	0	0	1	1	1	0	0	0	0	0	0	0	0	0	0		–		0	0	1	1	0
	0		0	0	1	1	1	0	0	1	1	1	0	0	0	0	0	0	0	0	0	0		–		0	0	1	1	0
0	0	1	0	0	1	1	1	0	0	1	1	1	0	0	0	0	0	0	0	0	0	0		–		0	0	1	1	0
1	0	1	0	0	0	0	0	0	0	0	0	0	0	0	1	1	1	0	0	1	0	0	1	–	1	1	1	0	1	0
1	0	1	0	0	0	0	0	0	0	0	0	0	0	0	1	1	0	0	0	0	0	0	1	–	0	1	1	0	1	0
1	0	1	0	0	1	1	1	0	0	1	1	1	0	0	0	0	0	0	0	0	0	0	1	–	1	0	0	1	1	0
0	0	1	0	0	1	1	1	0	0	1	1	1	0	0	0	0	0	0	0	0	0	0	0	–	1	0	0	1	1	0
1	1	1	1	1	1	1	1	0	0	1	1	1	0	0	0	0	0	0	0	0	0	1	1		1	0	0	1	1	0
0	0	1	0	0	0	0	0	0	0	0	0	0	0	0	1	1	1	0	1	0	0	0	0	–		0	0	1	1	0
1		1	0	0	0	0	0	0	0	0	0	0	0	0	1	1	1	0	1	0	0	0	0	–	1	0	0	1	1	0
1	1	1	1	1	0	0	0	0	0	0	0	0	0	0	1	1	1	0	1	1	0	0	0	–	1	0	–	–	1	0
0	0	1	0	0	0	0	0	0	0	0	0	0	0	0	1	1	1	0	1	1	0	0	0	–	1	0	–	–	1	0
0			0	0	0	0	0	0	0	0	0	0	0	0	0	0	0	0	0	0	0	0			1	0	0	1	1	0
1	0	1	0	0	0	0	0	0	0	0	0	0	0	0	1	1	1	0	1	1	0	0	0	–	1	0	–	–	1	0
0	1	1	0	0	0	0	0	0	0	0	0	0	0	0	1	1	0	0	0	1	0	0	0	–	1	0	–	–	1	0
			0	0	0	0	0	0	0	0	0	0	0	0	1	1	1	0	0	1	0	0	0	–		0	–	–	1	0
			0	0	1			0	0							0	0	0	0	0	0	0	0	–		0	–	–	1	1
1	0	1	0	0	0	0	0	0	0	0	0	0	0	0	1	1	1	0	1	1	0	0	0	–	1	0	0	1	1	0
1	0	1	0	0	0	0	0	0	0	0	0	0	0	0	1	1	0	0	0		0	0	0	–	1	0	0	1	1	0

	T_0	T_1	T_2	$T_{2\frac{1}{2}}$	T_3	$T_{3\frac{1}{2}}$	T_4	T_5	Urysohn	Semiregular	Regular	Completely Regular	Normal	Completely Normal	Perfectly Normal	Compact	σ-Compact	Lindelöf	Countably Compact	Sequentially Compact	Weak. Count. Compact	Pseudocompact	Locally Compact	Strong Loc. Compact	σ-Locally Compact	Separable	Second Countable	First Countable	Count. Chain Cond.	Paracompact
95	1	1	1	1	1	1	1	1	1	1	1	1	1	1	1	1	1	1	1	1	1	1	1	1	1	1	0	1	1	1
96	1	1	1	1	0	0	0	0	1	0	0	0	0	0	0	0	0	1	0	0	0		0	0	0	1	0	1	1	0
97	1	1	1	1	1	1	1	1	1	1	1	1	1	1	0	1	1	1	1	1	1	1	1	1	1	0	0	1	0	1
98	1	1	1	1	1	1	1	1	1	1	1	1	1	1	1	0	1	1	0	0	0	0	0	0	0	1	0	0	1	1
99	1	1	0	0	0	0	0	0	0	0	0	0	0	0	0	1	1	1	1	1	1	1	1	1	1	1	0	0	1	1
100	1	1	1	0	0	0	0	0	0	1	0	0	0	0	0	0	1	1	0	0	0	1	0	0	0	1	1	1	1	0
101	1	1	1	1	1	1	1	0	1	1	1	1	1	0	0	1	1	1	1	1	1	1	1	1	1	0	0	0	0	1
102	1	1	1	1	1	1	1	1	1	1	1	1	1	1	1	0	0	1	0	0	0	0	0	0	0	1	1	1	1	1
103	1	1	1	1	1	1	0	0	1	1	1	1	0	0	0	0	0	0	0	0	0	0	0	0	0	–	0	0	1	0
104	1	1	1	1	1	1	1	1	1	1	1	1	1	1	1	0	0	0	0	0	0	0	0	0	0	0	0	1	0	1
105	1	1	1	1	1	1	1	0	1	1	1	1	1	0	0	1	1	1	1	0	1	1	1	1	1	1	0	0	1	1
106	1	1	1	1	1	1	0	0	1	1	1	1	0	0	0	0	0	0	1	0	1	1	1	1	0	0	0	0	0	0
107	1	1	1	1	1	1	1	0	1	1	1	1	1	0	0	1	1	1	1	1	1	1	1	1	1	1	0	1	1	1
108	1	1	1	1	1	1	1	1	1	1	1	1	1	1	1	0	0	1	0	0	0	0	0	0	0	1	1	1	1	1
109	1	1	1	1	1	1			1	1	1	1				0	0	0	0	0	0	0	0	0	0	0	0	0	0	
110																														
111	1	1	1	1	1	1	1	0	1	1	1	1	1	0	0	1	1	1	1	0	1	1	1	1	1	1	0	0	1	1
112	1	1	1	1	1	1			1	1	1	1				0	0	0	1		1	1			0	1	0		1	0
113	1	1	1	1	0	0	0	0	1	0	0	0	0	0	0	0	0	0	0	0	0	1	0	0	0	1	0	0	1	0
114	1	1	1	1	1	1	1	1	1	1	1	1	1	1	1	0	1	1	0	0	0	0	0	0	0	1	0	0	1	1
115	1	1	1	1	1	1	1	1	1	1	1	1	1	1	1	0	1	1	0	0	0	0	0	0	0	1	1	1	1	1
116	1	1	1	1	1	1	1	1	1	1	1	1	1	1	1	0	1	1	0	0	0	0	0	0	0	1	1	1	1	1
117	1	1	1	1	1	1	1	1	1	1	1	1	1	1	1	1	1	1	1	1	1	1	1	1	1	1	1	1	1	1
118	1	1	1	1	1	1	1	1	1	1	1	1	1	1	1	1	1	1	1	1	1	1	1	1	1	1	1	1	1	1
119	1	1	1	1	1	1	1	1	1	1	1	1	1	1	1	0	1	1	0	0	0	0	1	1	1	1	1	1	1	1
120	1	1	1	1	1	1	1	1	1	1	1	1	1	1	1	1	1	1	1	1	1	1	1	1	1	1	1	1	1	1
121	1	0	0	0	0	0	1	1	0	0	0	0	0	0	0	1	1	1	1		1	1	1	1	1	1			1	1
122	1	1	1	1	1	1	1	1	1	1	1	1	1	1	1	0	1	1	0	0	0	0	1	1	1	1	1	1	1	1
123	1	1	1	1	1	1	1	1	1	1	1	1	1	1	1	0	1	1	0	0	0	0	1	1	1	1	1	1	1	1
124																														
125	1	1	1		0	0	0	0	0		0	0	0	0	0	0	1	1	0	0	0	1	0	0	0	1			1	0
126	1	1	1	1	0	0	0	0	0	0	0	0	0	0	0	0	1	1	0	0	0	1	0	0	0	1	1	1	1	0
127	1	1	1	1	0	0	0	0	1		0	0	0	0	0	0	1	1	0	0	0		0	0	0	1	1	1	1	0
128	1	1	1	1	1	1	1	1	1	1	1	1	1	1	1	0	0	1	0	0	0	0	0	0	0	1	1	1	1	1
129	1	1	1	1	1	1	1	1	1	1	1	1	1	1	1	0	0	1	0	0	0	0	0	0	0	1	1	1	1	1
130	1	1	1	1	1	1	1	1	1	1	1	1	1	1	1	1	1	1	1	1	1	1	1	1	1	1	1	1	1	1

GENERAL REFERENCE CHART (*continued*)

Metacompact	Count. Paracompact	Count. Metacompact	Fully Normal	Fully T_4	Connected	Path Connected	Arc Connected	Hyperconnected	Ultraconnected	Locally Connected	Local Path Connected	Local Arc Connected	Biconnected	Has Dispersion Point	Totally Path Disc.	Totally Disconnected	Totally Separated	Extremally Disc.	Zero Dimensional	Scattered	Discrete	Metrizable	σ-Locally Finite Base	Topol. Complete	Second Category	Countable	Card Less Than c	Card = c	Card Not Exc. 2^c	Strongly Connected
1	1	1	1	1	0	0	0	0	0	0	0	0	0	0	1	1	1	0	1	0	0	0	0	–	1	0	0	1	1	0
1	0	1	0	0	0	0	0	0	0	0	0	0	0	0	1	1	1	0	0	0	0	0		–		0	0	1	1	0
1	1	1	1	1	0	0	0	0	0	0	0	0	0	0	0	0	0	0	0	0	0	0	0	–	1	0	0	1	1	0
1	1	1	1	1	0	0	0	0	0	0	0	0	0	0	1	1	1	0	1	1	0	0	0	–	1	1	1	0	1	0
1	1	1	0	0	0	0	0	0	0	0	0	0	0	0	1	1	0	0	0	1	0	0	1	–	1	1	1	0	1	0
1	0	1	0	0	0	0	0	0	0	0	0	0	0	0	1		0	0	0		0	0	1	–	1	1	1	0	1	0
1	1	1	1	1	1	1	1	0	0	0	0	0	0	0	0	0	0	0	0	0	0	0	0	–	1	0	0	1	1	0
1	1	1	1	1	0	0	0	0	0	0	0	0	0	0	1	1	1	0	1	0	0	1	1	1	1	0	0	1	1	0
			0	0	0	0	0	0	0	0	0	0	0	0	1	1	1	0	1	0	0	0	0	–		0	–	–	–	0
1	1	1	1	1	0	0	0	0	0	0	0	0	0	0	1	1	1	0	1	0	0	1	1	1	1	0	0	1	1	0
1	1	1	1	1	1	1	1	0	0	1	1	1	0	0	0	0	0	0	0	0	0	0	0	–	1	0	0	0	1	0
0	1	1	0	0	0	0	0	0	0	0	0	0	0	0	0	0	0	0	0	0	0	0	0	–	1	0	0	0	1	0
1	1	1	1	1	1	1	1	0	0	1	1	1	0	0	0	0	0	0	0	0	0	0	0	–	1	0	0	1	1	0
1	1	1	1	1	1	1	1	0	0	1	1	1			0	0	0	0	0	0	0	1	1	1	1	0	0	1	1	0
					0	0	0	0	0	0	0	0	0	0	0	0	0	0	0	0	0	0	0	–	1	0	0	1	1	0
1	1	1	1	1	0	0	0	0	0	0	0	0	0	0	1	1	1	1	1	0	0	0	0	–	1	0	0	0	1	0
0	1	1	0	0				0	0						1	1	1		1		0	0	0	–	1	0			1	0
0			0	0	0	0	0	0	0	0	0	0	0	0	1	1	1	1	0	1	0	0		–		0			1	0
1	1	1	1	1	0	0	0	0	0	0	0	0	0	0	1	1	1	1	1	1	0	0	0	–	1	1	1	0	1	0
1	1	1	1	1	0	0	0	0	0	0	0	0	0	0	0	0	0	0	0	0	0	1	1	1	1	0	0	1	1	0
1	1	1	1	1	1	0	0	0	0	0	0	0	0	0	0	0	0	0	0	0	0	1	1		1	0	0	1	1	0
1	1	1	1	1	1	0	0	0	0	0	0	0	0	0	0	0	0	0	0	0	0	1	1	1	1	0	0	1	1	0
1	1	1	1	1	1	1	1	0	0	0	0	0	0	0	0	0	0	0	0	0	0	1	1	1	1	0	0	1	1	0
1	1	1	1	1	1	0	0	0	0	0	0	0	0	0	0	0	0	0	0	0	0	1	1		1	0	0	1	1	0
1	1	1	1	1	1	1	1	0	0	0	0	0	0	0	0	0	0	0	0	0	0	1	1	1	1	0	0	1	1	0
1	1	1	0		1	1	0	0	1	0	0	0	0	0	0	0	0	0	0		0	0		–	0	1	1	0	1	1
1	1	1	1	1	1	0	0	0	0	0	0	0	0	0	0	0	0	0	0	0	0	1	1	1	1	0	0	1	1	0
1	1	1	1	1	1	0	0	0	0	0	0	0	0	0	0	0	0	0	0	0	0	1	1		1	0	0	1	1	0
1	0	1	0	0	1	0	0	0	0		0	0			1	0	0	0	0	0	0	0		–		1	1	0	1	1
1	0	1	0	0	1	0	0	0	0	0	0	0	1	1	1	0	0	0	0	0	0	0	1	–	0	1	1	0	1	1
1	0	1	0	0	0	0	0	0	0	0	0	0	0	0	1	1	1	0	0	0	0	0	1	–	0	1	1	0	1	0
1	1	1	1	1	1	0	0	0	0	0	0	0	1	1		0	0	0	0	0	0	1	1			0	0	1	1	0
1	1	1	1	1	0	0	0	0	0	0	0	0	0	0	1	1	0	0	0	0	0	1	1			0	0	1	1	0
1	1	1	1	1	1	0	0	0	0		0	0			1	0	0	0	0	0	0	1	1	1	1	0	0	1	1	0

	T_0	T_1	T_2	$T_{2\frac{1}{2}}$	T_3	$T_{3\frac{1}{2}}$	T_4	T_5	Urysohn	Semiregular	Regular	Completely Regular	Normal	Completely Normal	Perfectly Normal	Compact	σ-Compact	Lindelöf	Countably Compact	Sequentially Compact	Weak. Count. Compact	Pseudocompact	Locally Compact	Strong Loc. Compact	σ-Locally Compact	Separable	Second Countable	First Countable	Count. Chain Cond.	Paracompact
131	1	1	1	1	1	1	1	1	1	1	1	1	1	1	1			1								1	1	1	1	1
132	1	1	1	1	1	1	1	1	1	1	1	1	1	1	1	0	0	0	0	0	0	0	0	0	0	0	0	1	0	1
133	1	1	1	1	0	0	0	0	1	0	0	0	0	0	0	0	0	0	0	0	0	0	0	0	0	1	0	1	1	0
134																														
135	1	1	1	1	1	1	1	1	1	1	1	1	1	1	1	0	1	1	0	0	0	0	1	1	1	1	1	1	1	1
136	1	1	1	1	1	1	1	1	1	1	1	1	1	1	1	0	0	1	0	0	0	0	0	0	0	1	1	1	1	1
137																														
138																														
139	1	1	1	1	1	1	1	1	1	1	1	1	1	1	1	0	0	0	0	0	0	0	0	0	0	0	0	1	0	1
140	1	1	1	1	1	1	1	1	1	1	1	1	1	1	1	0	0	0	0	0	0	0	0	0	0	0	0	1	0	1
141	1	1	1	1	1	1	1	1	1	1	1	1	1	1	1	0	0	0	0	0	0	0	0	0	0	0	0	0	0	1
142	1	1	1	1	1	1	1	1	1	1	1	1	1	1	0	0	0	0							0	0	0	0	0	0
143	1	1	1	1	1	1	1	1	1	1	1	1	1	1	0	0	0	0	0	0	0	0			0	0	0	0	0	0

GENERAL REFERENCE CHART (continued)

Metacompact	Count. Paracompact	Count. Metacompact	Fully Normal	Fully T_4	Connected	Path Connected	Arc Connected	Hyperconnected	Ultraconnected	Locally Connected	Local Path Connected	Local Arc Connected	Biconnected	Has Dispersion Point	Totally Path Disc.	Totally Disconnected	Totally Separated	Extremally Disc.	Zero Dimensional	Scattered	Discrete	Metrizable	σ-Locally Finite Base	Topol. Complete	Second Category	Countable	Card Less Than c	Card = c	Card Not Exc. 2^c	Strongly Connected
1	1	1	1	1	1			0	0				1	0		0	0	0	0	0	0	1	1			0	0		1	0
1	1	1	1	1	1	1	1	0	0	1	1	1	0	0	0	0	0	0	0	0	0	1	1	1	1	0	0	1	1	0
					1																									
1	1	1	1	1	0	0	0	0	0	1	1	1	0	0	1	1	1	1	1	1	1	1	1	1	1	1	1	0	1	0
1	1	1	1	1	0	0	0	0	0	0	0	0	0	0	1	1	1	0		0	0	1	1			0	0	1	1	0
1	1	1	1	1	0	0	0	0	0	0	0	0	0	0	1	1	1	0	1	1	0	1	1	1	1	0	0	1	1	0
1	1	1	1	1	1	1	1	0	0	1	1	1	0	0	0	0	0	0	0	0	0	1	1	1	1	0	0	1	1	0
1	1	1	1	1	1	1	1	0	0	1	1	1	0	0	0	0	0	0	0	0	0	0	0	–	1	0	0	1	1	0
0			0	0	0	0	0	0	0													0	0	–	1	0	0	0	0	0
1	1	1	0	0	0	0	0	0	0													0	0	–		0	0	0	0	0

PROBLEMS

SECTION 1

1. If $\{A_i\}_{i=1}^{\infty}$ is a countably infinite collection of subsets of a topological space, show that $\bigcup_{i=1}^{\infty} \bar{A}_i \subset \overline{\bigcup_{i=1}^{\infty} A_i}$.
2. True or false: $\overline{A \cap B} = \bar{A} \cap \bar{B}$.
3. Show that the complement of an F_σ set is a G_δ set, and conversely.
4. Show that any space with an open point must be second category. What is the smallest second category space?
5. Show that if for $i = 1, \ldots, n$, C_i is a closed subset of a topological space X, and $f\colon X \to Y$ is continuous on C_i for all i, then f is continuous on $\bigcup_{i=1}^{n} C_i$. Show that this result does not hold if one considers infinitely many closed sets.
6. Show that a filter F on a set X is an ultrafilter if and only if for every two disjoint subsets A and B of X such that $A \cup B \in F$, either $A \in F$ or $B \in F$.

SECTION 2

7. Show that a space is T_1 if and only if every point is closed.
8. Show that a space is T_2 if and only if every point is the intersection of its closed neighborhoods.
9. Show that a space is T_3 if and only if every open set contains a closed neighborhood of each of its points.

10. Show directly that every second countable regular space is completely normal. (Do not use any metrization theorems.)
11. Show that ΠX_α is completely regular if each X_α is completely regular.
12. Show that every Urysohn space is completely Hausdorff.

Section 3

13. Show that every separable space satisfies the countable chain condition.
14. Prove the following generalization of the Tietze extension theorem: any real-valued continuous function on a closed subset X of a normal space Y may be extended continuously to all of Y.
15. Show that every fully normal space is normal.
16. Show that disjoint compact subsets of a Hausdorff space have disjoint neighborhoods.
17. Show that every paracompact Hausdorff space is normal.
18. Show that every σ-locally compact Hausdorff space is normal.
19. Show that every locally compact Hausdorff space is completely regular.
20. Show that every Lindelöf T_3 space is paracompact.
21. Show that every second countable T_3 space is both Lindelöf and T_5.
22. Prove Tychonoff's theorem: the product of an arbitrary family of topological spaces is compact iff each factor space is compact.
23. Is the product of second category spaces always second category?
24. Prove that every open subspace of a separable space is separable
25. Show that the countable Cartesian product of separable spaces is separable.

Section 4

26. Show that the following are equivalent:
    (i) X has no nontrivial separation
    (ii) X has no nontrivial subsets which are both open and closed.
27. Show that the union of any family of connected sets with a nonempty intersection is connected.
28. Show that if a space has just one quasicomponent, it must be connected.
29. Show that every quasicomponent in a locally connected space is connected.
30. Show that every countable T_1 space is totally path disconnected.
31. Show that every zero dimensional space is $T_{3\frac{1}{2}}$.

SECTION 5

32. Show that every metric space is perfectly normal.
33. Show that a metric space is compact if and only if it is complete in every equivalent metric.
34. Show that every second countable space has a σ-locally finite base.

COUNTEREXAMPLES

35. Show that the indiscrete topology on a set (Example 4) is arc connected iff the set is uncountable.
36. Show that the uncountable particular point topology (Example 10) does not have a σ-locally finite base.
37. Show that the uncountable excluded point topology (Example 15) has a σ-locally finite base.
38. Show that the either-or topology (Example 17) has a σ-locally finite base.
39. Prove that the finite complement topology on an uncountable set (Example 19) is second category.
40. Show that the countable complement topology (Example 20) is not path connected.
41. Show that the countable complement topology (Example 20) does not have a σ-locally finite base.
42. Show that the countable complement topology (Example 20) is second category by showing that a set is nowhere dense if and only if it is countable.
43. Show that the compact complement topology (Example 22) is second countable.
44. Show that the compact complement topology (Example 22) is not second category.
45. Countable Fort space (Example 23) is metrizable since it is regular and second countable. Find a metric which gives this topology.
46. Show that Fortissimo space (Example 25) does not satisfy the countable chain condition and thus is not second countable.
47. Prove that the real line R (Example 28) is a complete metric space.
48. Show that the rational numbers are dense in the real line (Example 28).
49. Show that a subset of Euclidean n-space (Example 28.9) is compact iff it is closed and bounded.
50. What can be said about the cardinality of connected subsets of R^n (Example 28)?
51. Show that the uniformity $\{S_{ab}\}_{a,b\in R}$ where $S_{ab} = \{(x,y)|x,y < b$ or $x,y > a\}$ is not the usual metric uniformity for the real line (Example 28) but still gives the Euclidean topology.

52. Prove that the Cantor set (Example 29) is zero dimensional.
53. Show that the metric d of Example 30.5 is indeed a metric for the real numbers.
54. Show that the rational numbers with the Euclidean topology (Example 30) are not topologically complete.
55. The set of irrationals in [0,1] (Example 30) is topologically complete but not compact; thus it cannot be totally bounded. Show this directly from the definition of totally bounded.
56. Show, without using the concept of compactness that (0,1) is not homeomorphic to [0,1] (Example 32.7).
57. Show that no homeomorphism of R onto itself can map $A = \{0\} \cup [1,2] \cup \{3\}$ onto $B = [0,1] \cup \{2\} \cup \{3\}$, even though A is homeomorphic to B (Example 32.8).
58. Show that the one point compactification of the irrationals is second category, but not first countable (Example 34).
59. Show that the one point compactification of the irrationals is not arc connected (Example 34).
60. Note that the one point compactification of the irrationals (Example 34) is of course locally compact. Why should this be considered artificial?
61. Show that the Fréchet product metric (Example 37.7) does indeed give the right topology for Hilbert space.
62. Show that every separable metric space may be imbedded in Fréchet space (Example 37).
63. Show that every connected order topology (Example 39) is locally connected. More generally, show that any connected topology on a linearly ordered set is locally connected provided it has a basis of convex sets.
64. Ordinal spaces for countable ordinals (Examples 40 and 41) are metrizable. Find appropriate metrics.
65. Show that open ordinal space $[0,\Gamma)$ for $\Gamma < \Omega$ (Example 40) is topologically complete.
66. Give as many different reasons as possible why closed uncountable ordinal space (Example 43) is not metrizable.
67. The extended long line (Example 46) is not path connected since no path can join any point to Ω. Prove this.
68. There is an obvious definition of $\sin 2\pi x$ for every $x \in L^*$ the extended long line (Example 46). Why is this function not continuous?
69. Show that the altered long line (Example 47) is not locally compact.
70. Show that the altered long line (Example 47) does not have a σ-locally finite base by showing that this property is preserved in open subspaces.

71. Prove that the lexicographic ordering on the unit square (Example 48) does not yield a perfectly normal topology.
72. Prove that the unit square with the lexicographic ordering topology (Example 48) is indeed first countable.
73. Show that every right order topology (Example 49) is locally compact.
74. Show that the right half open interval topology (Example 51) is neither locally compact nor second category.
75. Show that the right half open interval topology (Example 51) is perfectly normal.
76. Show that the nested interval topology (Example 52) is not second category.
77. Show that the overlapping interval topology (Example 53) is second category.
78. Find an infinite subset of the interlocking interval topology (Example 54) which does not have a limit point.
79. Show that the interlocking interval topology (Example 54) is not strongly locally compact.
80. Show that the interlocking interval topology (Example 54) is neither second countable, scattered, nor biconnected.
81. Show that the prime ideal topology (Example 56) is second category.
82. Show that the divisor topology (Example 57) is not fully T_4.
83. Show that with the divisor topology (Example 57), the positive integers are weakly countably compact but not countably metacompact.
84. The evenly spaced integer topology (Example 58) is metrizable. Find a metric which yields this topology on the integers.
85. Show that the integers Z with the p-adic topology (Example 59) are not extremally disconnected.
86. Show that the relatively prime integer topology (Example 60) is not biconnected. *Hint:* first show that the prime integer topology (Example 61) is not biconnected.
87. Prove the assertion (Example 63.7) that a subset of the countable complement extension topology is compact iff it is finite.
88. Show that the countable complement extension topology (Example 63) is neither pseudocompact nor metacompact. From what other property of this space can you then determine immediately that it is not countably metacompact?
89. Prove that the countable complement extension topology (Example 63) satisfies the countable chain condition.
90. Show that Smirnov's deleted sequence topology (Example 64) is second countable.
91. Show that Smirnov's deleted sequence topology (Example 64) is connected, but neither path connected nor locally connected.

92. Show that in the rational sequence topology (Example 65) every subset is a G_δ set.
93. Show that the rational sequence topology (Example 65) is not paracompact.
94. Show that the rational sequence topology (Example 65) is not normal.
95. Discuss the rational sequence topology (Example 65) with regard to whether it is ever countably paracompact for any choice of sequences.
96. Show that both the indiscrete extensions of R (Examples 66 and 67) as well as the pointed extensions of R (Examples 68 and 69) fail to be semiregular.
97. Show that the indiscrete rational extension of R (Example 66) is metacompact.
98. Show that neither discrete extension of R (Examples 70 and 71) is σ-compact.
99. Show that the discrete rational extension of R (Example 70) is zero dimensional, but neither scattered nor extremally disconnected.
100. Show that the discrete rational extension of R (Example 70), with the metric given in 70.3, is not complete. Is this space topologically complete?
101. Show that the double origin topology (Example 74) is σ-compact.
102. Show that the irrational slope topology (Example 75) is not second category.
103. Show the deleted diameter and radius topologies (Examples 76 and 77) are arc connected. *Hint:* consider paths which contain no horizontal segments at all.
104. Show that although neither the deleted diameter nor the deleted radius topologies (Examples 76 and 77) is second countable, the deleted radius topology is Lindelöf.
105. Is the deleted radius topology (Example 77) metacompact?
106. Show that the half-disc topology (Example 78) is arc connected.
107. Show that the irregular lattice topology (Example 79) is second category.
108. Justify the global and local compactness properties of Arens square (Example 80).
109. Show that the space developed in Example 82.9 from Niemytzki's tangent disc topology is normal.
110. Prove that Niemytzki's tangent disc topology (Example 82) is neither Lindelöf nor σ-compact.
111. Prove that Niemytzki's tangent disc topology (Example 82) is arc connected.

112. Show that the metrizable tangent disc topology (Example 83) is arc connected.
113. Show that Sorgenfrey's half open square topology (Example 84) is countably metacompact.
114. Show that Michael's product topology (Example 85) is zero dimensional, but neither scattered nor extremally disconnected.
115. Show that Michael's product topology (Example 85) is first countable.
116. Show that Michael's product topology (Example 85) is second category.
117. Show that the deleted Tychnoff plank (Example 87) is locally compact but not Lindelöf.
118. Show that the Alexandroff Plank (Example 88) is neither σ-compact nor Lindelöf.
119. Verify that the Dieudonne plank (Example 89) does not satisfy any of the global or local compactness properties.
120. Prove the assertion that $\psi^{-1}(N \cap X) \subset N$ which appears in the construction of Hewitt's condensed corkscrew (Example 92).
121. Prove that the strong parallel line topology (Example 96) is neither sequentially compact nor paracompact, but is metacompact.
122. Show that the concentric circles topology (Example 97) is not perfectly normal.
123. Show that the minimal Hausdorff topology (Example 100) is pseudocompact.
124. Show that the Alexandroff square (Example 101) is neither perfectly normal nor separable.
125. A metric space is compact iff it is complete in every metric. Z^Z (Example 102) is not compact though we describe a metric in which it is complete. Find a metric in which it is not complete.
126. Show that the uncountable product of copies of Z^+ (Example 103) is neither Lindelöf nor extremally disconnected.
127. Show that $[0,\Omega) \times I^I$ (Example 106) is not normal.
128. Show that the subspace Y of Helly space consisting of continuous piecewise linear functions, which take rational values on the diadic rationals (Example 107.3), is dense in Helly space. Further show that Helly space is not completely normal.
129. Show that the box product topology on R^ω (Example 109) gives a space which is not Lindelöf.
130. Show that the Stone-Čech compactification of the integers (Example 111) is not first countable.

131. Novak space (Example 112) is clearly not compact. Find an open cover with no finite subcover.
132. Show that the strong ultrafilter topology (Example 113) is an expansion of the Stone-Čech compactification of the positive integers.
133. Show that the strong ultrafilter topology (Example 113) is neither locally compact nor first countable.
134. Show that the single ultrafilter topology (Example 114) is perfectly normal and paracompact, but not locally compact.
135. Show that the integer broom (Example 121) is T_5.
136. The construction of Bernstein's connected sets (Example 124) assumes that the number of closed connected subsets of R^n is c, the power of the continuum. Prove this.
137. Prove that the pseudo-arc (Example 130) is nonempty.
138. Prove that the wheel without its hub (Example 132) is locally arc connected.
139. Prove that the wheel without its hub (Example 132) is not Lindelöf.
140. Show that the wheel without its hub (Example 132) is not locally compact.
141. Show that the wheel without its hub (Example 132) is topologically complete, though not complete in the given metric.
142. Be heroic. Verify that the function $\sigma(x,y)$ in Example 134.8 is indeed a metric.
143. Verify that Sierpinski's metric (Example 135) on a countable set satisfies the triangle inequality (axiom M_2 for a metric).
144. Show that Duncan's space (Example 136) is not complete in the given metric. Is there a metric in which this space is complete?
145. Determine whether Duncan's space (Example 136) is zero dimensional.
146. Fill in the missing details in the construction of the Cauchy completion of a metric space (Example 137.1).
147. Show that the plane with the post office metric (Example 139) is complete. Since it is not compact, it is not complete in every equivalent metric. Find a metric for this space which is not complete.
148. Show that the radial metric (Example 140) really is a metric and that it yields a complete metric space.
149. Show that the plane with the radial metric topology (Example 140) is not locally compact.

NOTES

Part I: Basic Definitions

SECTION 1. GENERAL INTRODUCTION

1. In the definition of a topological space, condition O_3 is actually redundant since the union of an empty family of sets is empty, and the intersection of an empty family of subsets of a set X is X itself.

2. With the abbreviations introduced in Example 32.9 we can explicitly represent the semigroup of sets formed by complementation and closure (Table 2). The inclusion relations between these fourteen sets can be summarized by

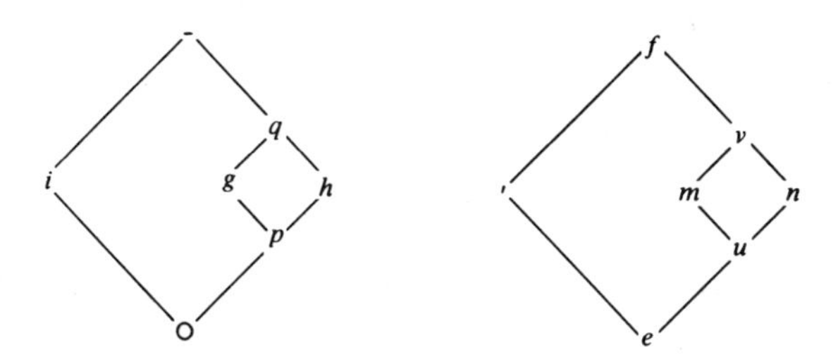

where larger sets are above smaller ones.

3. The Tychonoff topology on the set $\prod_{\alpha \in A} X_\alpha$ is characterized by the following universal property: if Z is any topological space, and if $f_\alpha\colon Z \to X_\alpha$ are continuous there exists a unique continuous function $f\colon Z \to \prod_{\alpha \in A} X_\alpha$ such that $\pi_\alpha \circ f = f_\alpha$. Indeed $\prod_{\alpha \in A} X_\alpha$ with the Tychonoff topology is the product in the category of topological spaces and continuous mappings. Likewise the quotient space X/R is universal with

Table 2

OPERATIONS TABLE

i = identity	e = –′	m = fo
′ = complement	f = o′	n = e–
o = interior	g = o–	p = go
– = closure	h = –o	q = h–
u = q′		v = p′

	i	′	o	–	e	f	g	h	m	n	p	q	v	u
i	i	′	o	–	e	f	g	h	m	n	p	q	v	u
′	′	i	e	f	o	–	n	m	h	g	u	v	q	p
o	o	f	o	g	m	f	g	p	m	v	p	q	v	m
–	–	e	h	–	e	n	q	h	u	n	h	q	n	u
e	e	–	e	n	h	–	n	u	h	q	u	n	q	h
f	f	o	m	f	o	q	v	m	p	g	m	v	g	p
g	g	m	p	g	m	v	g	p	m	v	p	g	v	m
h	h	n	h	q	u	n	q	h	u	n	h	q	n	u
m	m	g	m	v	p	g	v	m	p	g	m	v	p	g
n	n	h	u	n	h	q	n	u	h	q	u	n	q	h
p	p	v	p	g	m	v	g	p	m	v	p	g	v	m
q	q	u	h	q	u	n	q	h	u	n	h	q	n	u
v	v	p	m	v	p	g	v	m	p	g	m	v	g	p
u	u	q	u	n	h	q	n	u	h	q	u	n	q	h

respect to continuous functions $f\colon X \to Y$ such that the relation defined by f on X extends R. That is, if $\rho\colon X \to X/R$ is the projection and $f\colon X \to Y$ defines (by $x \sim x'$ if $f(x) = f(x')$) a relation which extends R then there exists a unique continuous function $f'\colon X/R \to Y$ such that $f' \circ \rho = f$. By the universality of ρ this condition characterizes X/R. Finally, the topological sum is characterized dually to the product by the inclusion functions of the summands, and thus is the sum or coproduct in the category of topological spaces and continuous functions.

4. When dealing with identification topologies the concept of a saturated set is often helpful. If $f\colon X \to Y$, and if $A \subset X$, A is called a **saturated** subset if $A = f^{-1}(B)$ for some B in Y, that is, A is the complete in-

verse image of some subset of Y. The map $\rho: X \to X/R$ is not generally open but it does take saturated open sets to open sets.

5. It should be noted that the existence of any nonprincipal ultrafilters depends on the axiom of choice for the construction of ultrafilters uses Zorn's lemma to produce at least one maximal ultrafilter containing a given filter. In Example 111, we construct many distinct ultrafilters on a countable set; in fact, we construct 2^c ultrafilters of which only $\aleph_0$ can be principal.

Section 2. SEPARATION AXIOMS

1. Certain general constructions dealing with the separation axioms are worth special note because of their generality and effectiveness. Many others of more special applicability can be found among the examples. If one doubles the points of a space (technically this involves taking the product of the space with the two point indiscrete space) the resulting space is no longer T_0, T_1, or T_2, but clearly the new space satisfies the same higher T_i-axioms as did the original space. More generally since for $i \leq 3\frac{1}{2}$, the product of two spaces is T_i iff each space is T_i, we may use products to destroy certain selected T_i properties.
2. Properties T_4 and T_5 are often satisfied vacuously if the space X has no disjoint closed sets. The open extension topology (Example 16) is a general construction which accomplishes this.
3. The results that a space is T_5 iff every subspace is T_4 and that every perfectly normal space is completely normal may be found in Gaal [42].

Section 3. COMPACTNESS

1. A proof of Alexander's subbasis theorem can be found in Gaal [42], p. 146.
2. It should be noted that our definition of countably compact requires that infinite sets have ω-accumulation points, while most authors give a definition of countably compact which is equivalent to our definition of weakly countably compact. Since in a T_1 space every limit point is an ω-accumulation point it is clear that the two different definitions coincide in a T_1 space. Hence the theorems which correspond to the equivalences of CC_1, CC_2, CC_3, and CC_4 usually involve the assumption that the space involved is T_1. In particular the proof due to Arens and Dugundji [35], p. 229, that a space is compact iff it is both countably compact and metacompact does not need the assumption that the space

is T_1 if one defines countable compactness using ω-accumulation points. Finally one should note that doubling the points of any space renders it weakly countably compact, for then every subset has a limit point, namely, the twin of one of its points. Note that the space is no longer T_0 and in fact is still not countably compact if it were not so originally.

3. The Venn diagram which relates the countability axioms and compactness omits the concept of separability since there are general means available to render each example separable or nonseparable, as desired. To make any space separable without affecting any of the other properties involved in the Venn diagram one simply takes the closed extension of that space (Example 12). The new point is then a dense subset. Conversely, to render the space nonseparable it is sufficient to take the product of the space with uncountable Fort space (Example 24) which is compact and nonseparable. The result will then have exactly the compactness properties of the original space but will no longer be separable. There are other useful tricks for producing certain desired alterations. We have already observed that the product of any space with the two point indiscrete space effectively doubles the points of the original space thus rendering it non-Hausdorff and all that that entails. Direct sums of two spaces often have a different variety of properties than either of the summands. And finally, the methods of the indiscrete, pointed, and discrete extensions (Examples 66–71) are frequently useful in dealing with the higher separation axioms.

4. Further discussion of fully normal spaces together with proofs of the relation to normal and paracompact spaces can be found in Gaal [42]. Proofs of the T_2 and T_3 implications concerning the compactness properties can be found in Dugundji [35] and Gaal [42].

5. The product property lists separability as preserved under countable but not uncountable products. In fact, it is preserved under products of cardinality no greater than $2^{\aleph_0}$; this is proved in Dugundji [35], p. 175, and his proof is adapted in Example 103 to show that that product space is separable for $\lambda \leq 2^{\aleph_0}$. Dugundji also proves that every separable Hausdorff space has cardinality less than or equal to $2^{2^{\aleph_0}}$.

6. Many of the countability and compactness properties are cardinality (set theoretic) statements that in some way measure the size of a topological space. The general idea of cardinal functions in topology has now appeared as the title of a text by Juhász [60] in which many new and important cardinality conditions are studied.

7. Certain recent work has dealt with contrasts between separability and the Lindelöf property, properties which are equivalent in

metrizable spaces. The lexicographic square (Example 48) is an example of a non-separable Lindelöf space. On the other hand the Cantor tree (Example 14 in Part III)—an example of importance in metrization theory—is an example of a separable non-Lindelöf space. Moving still further into the realm of metrization theory, M. E. Rudin [103] used a Souslin line to construct recently an example of a normal non-Lindelöf space which is hereditarily separable.

Section 4. CONNECTEDNESS

1. The proof that local connectedness is preserved under certain continuous functions actually shows more. We observe that if the function maps saturated open sets (open sets which are complete inverse images of sets) to open sets then f preserves local connectedness. This condition is always satisfied if the image of f bears the identification topology. From this viewpoint, the given proof merely asserts that any Hausdorff image of a compact space bears the identification topology.

2. Countable spaces have some interesting connectivity properties. If X is countable and T_1 then it is not path connected for the inverse images of the points in the path would yield a decomposition of the closed unit interval into a denumerable number of closed disjoint subsets, a contradiction. If X is countable and connected it may not be Urysohn, for if it were Urysohn then there would be a nonconstant real-valued function on X; the image of this function must be countable, hence not connected. The inverse images of two components of the image will then separate the original space. Finally, and trivially, no countable space or finite space with more than one point is arc connected.

3. It should be noted that the three point space with the indiscrete topology is a biconnected space with no dispersion point. Miller's example (Example 131) is of interest because it is Hausdorff.

Section 5. METRIC SPACES

1. As in metric spaces the sets $B(x,\epsilon) = \{y \in X | d(x,y) < \epsilon\}$ form a basis for a topology whenever d is either a pseudometric or a quasimetric. The topology resulting from a pseudometric is not necessarily T_i for $i < 3$, but it is always T_3 and T_5 for the same reason that metric spaces are T_3 and T_5. For example, the indiscrete topology is given by pseudometric $d(x,y) = 0$ for all $x,y \in X$. Quasimetric spaces are discussed in Murdeshwar and Naimpally [85]. The compactness relations

for metric spaces also hold for pseudometric spaces with the exception that weak countable compactness need not imply countable compactness, for this result depends on the T_1 axiom.

2. It is shown in Pervin [92], p. 118 that in a totally bounded metric space every sequence contains a Cauchy subsequence. Thus if a totally bounded metric space is complete it is sequentially compact and hence compact.

3. The Baire category theorem and several equivalent formulations are presented in Pervin [92], pp. 127–128. Pervin also proves on p. 124 that all completions of a given metric space are isometric.

4. That regular second countable spaces are necessarily metrizable was proved by Urysohn [129] in 1924. In 1950 Bing [22], Nagata [87], and Smirnov [110] showed that a space is metrizable iff it is regular and has a σ-locally finite base. The search for further metrizability conditions continues, centering around the normal Moore space conjecture.

5. That a space is uniformizable iff it is $T_{3\frac{1}{2}}$ was proved in 1937 by Weil [136].

6. Niemytzki and Tychonoff [89] prove that a metric space is compact iff it is complete in each metric.

Part II: Examples

18. In 18.10 we prove X is not path connected by using the fact that the unit interval cannot be written as a countable disjoint union of closed sets. Since this result is used repeatedly to show certain spaces are not path connected, and since it is not usually proved in the standard texts, we prove it here.

Suppose $I = \bigcup_{i=1}^{\infty} C_i$ where $\{C_i\}$ is a family of disjoint closed sets; let $B = \bigcup \partial C_i = I - \bigcup C_i{}^\circ$. Then B is nowhere dense in I since each subinterval J of I contains an open subset L disjoint from B. This follows from the fact that J is of second category, so some C_k is dense in some open interval $L \subset J$; since C_k is closed $L \subset C_k{}^\circ$, so $L \cap B = \emptyset$. Since B is nowhere dense in I, every open interval U containing a point $x \in \partial C_j$ must intersect $B - \partial C_j$ for U, being a neighborhood of x, contains a point of $I - C_j$, say $u \in C_m{}^\circ$. Then if $U \cap B \cap \partial C_m = \emptyset$, $C_m{}^\circ \cap U$ is a nonempty open and relatively closed subset of U.

Now B itself is of second category (in itself) since it is a closed subset of I; thus some ∂C_k is dense in some nonempty open subset $U \cap B$ (where U is an open interval in I). Again, since ∂C_k is closed, this means that $\partial C_k \cap U = B \cap U$. But this is impossible, since if $U \cap \partial C_k \neq \emptyset$, then $U \cap (B - \partial C_k) \neq \emptyset$. This contradiction shows that I cannot be written as $\bigcup C_j$.

23. Fort [41] introduced this as an example of a Hausdorff space in which some points do not have a local basis of nested sets.

26. This more sophisticated type of Fort space is adapted from Arens

[11]. It is of interest particularly because it is a countable space which is not first countable.

28. This discussion of the Euclidean real line is somewhat incomplete in that it provides sanctuary for several logical circularities. First of all, we refrain from asserting that the real line is path connected for such a statement, though true, would be lacking significance since a path is defined to be the image of part of the real line. Furthermore, the fact that the real line is connected, and that the intervals $[a,b]$ are compact depends on the fact that the real line is complete, either in its order topology or in its metric. Since these completions are discussed later we chose to avoid using them in Example 28. A logically complete elementary discussion of the topological properties of the real line may be found in any introductory text on real analysis.

30. The complete metric on the irrationals is adapted from Greever [45], p. 110, where he proves a more general result due to Alexandroff [4] that every G_δ subspace of a complete metric space is topologically complete.

32. The sets in 32.9 are an explicit representation of the semigroup whose table is given earlier in these Notes.

36. The assertion that Hilbert space is homeomorphic to the countable infinite product of real lines was first proved by Anderson [8] in 1966. Anderson and Bing [9] provide an elementary, though lengthy proof of this result, together with a survey of related problems.

39. Example 132 shows that the converse of 39.9 is false: a topological space in which every point is a cut point need not be an order topology.

40. The proof in 40.12 that every continuous real-valued function on $[0,\Omega)$ is eventually constant is adapted from Dugundji [35], p. 81. This proof shows also that any continuous real-valued function on $[0,\Omega]$ is also eventually constant, though this fact can be proved more directly by observing that if $f(\Omega) = p$, then $f^{-1}(p) = f^{-1}(\bigcap B(p,1/n)) = \bigcap f^{-1}(B(p,1/n))$, a countable intersection of neighborhoods of Ω, which must contain some interval $(\alpha,\Omega]$.

45. Every metric space is perfectly normal, and if L were metrizable the proof in 45.4 would be unnecessary. But L is locally metrizable and this observation provides the idea for the proof as given.

L is normal but not paracompact, yet too large to be separable. Most examples of this type are non-separable, but M. E. Rudin has recently created an example [105] of a normal nonparacompact space which is separable.

47. This example was constructed by Alexandroff and Urysohn [5], pp. 71–72.

59. This topology is a special case of a topology for any commutative ring with unit that is usually called the A-adic topology where A is an ideal of the ring R. We take as a basis of neighborhoods of zero the sets A^n, the powers of the ideal A. We then take the set of cosets of these powers as the topology. R with this topology forms a topological group and thus is T_2 iff $\bigcap_{i=1}^{\infty} A^i = 0$. In the case that R is Hausdorff the function d defined by $d(r,s) = 2^{-k}$, where k is the largest power such that $r - s \in A^k$, defines a metric.

61. The proof that the prime integer topology σ is locally connected is due to Kirch [65].

74. Due to Alexandroff and Urysohn [5], p. 22.

75. Bing [18] introduced this as an example of a countable connected Hausdorff space; the first such example was given by Urysohn [127]. Bing's example, though connected, is not path connected, and the proof of this fact depends on the lemma proved above in the note for Example 18.

In [110] G. X. Ritter altered this example by enlarging neighborhoods to consist of a point in the plane together with triangles (shaded in the illustration on p. 93) on the x-axis. This space is then locally connected as well, giving an example of a connected and locally connected Hausdorff space that is simpler than Example 61. For other pathological examples of countable connected spaces see Example 126 and the related note.

80. Hewitt [51] credits Arens with constructing an example of this type; we present a modified version, and then a simplified version. Arens square is both semiregular and completely Hausdorff but not regular.

82. The method of argument in 82.7 centers on a subtle but very useful application of the Baire category theorem: if ϵ_x is a positive number for each real number x then at least one of the sets

$S_i = \{x \in R | \epsilon_x > 1/i\}$ is not nowhere dense, so there is some interval (a,b) and some i_0 where $\{x \in (a,b) | \epsilon_x > 1/i_0\}$ is dense in (a,b). This method of attack is used often in proofs concerning paracompactness and metacompactness.

83. This example is adapted from Bing [22], p. 182. He considers the case where the subset S is hereditarily G_δ, that is, where S and each of its subsets is G_δ. Now no set of cardinality $c = 2^{\aleph_0}$ can have this property since any such set has 2^c subsets, but there are only c G_δ sets. So the existence of an uncountable hereditarily G_δ set depends on the denial of the continuum hypothesis.

84. Sorgenfrey [112] used this example to show that paracompactness is not necessarily preserved even by finite products. More recently Przymusiński [94] has shown that under set theoretic conditions which are equivalent to the existence of nonmetrizable Moore spaces there exists a subset Y of S such that $Y \times Y \subset X$ is perfectly normal but still not paracompact.

85. This example is due to Michael [78].

88. This example is adapted from Alexandroff and Urysohn [5], p. 26.

89. This space was introduced by Dieudonne [33] in the same article in which he formulated the definition of a paracompact space.

90. The Tychonoff corkscrew was constructed by Arens and reported by Hewitt [51]. Greever [45] presents a lengthy exposition of the details of this example on pp. 77–79. Our presentation is a significant geometric rearrangement of the original, and our proofs rely more heavily on geometric intuition. Both Hewitt and Greever present the space as a cube with certain identifications along the edges; we have simply unfolded the cube into a corkscrew.

Recently a smaller (i.e., hereditarily separable) example of a regular but not completely regular space has been constructed by Ostaszewski [91].

92. This example, very complex yet very significant, was constructed by Hewitt [51] using a condensation process first described by Urysohn [127]. As in the previous example, we have relied heavily on a geometric analogy in order to present a clear description.

To appreciate the significance of this example, we should consider the relations between cardinality and connectedness. Any Urysohn space has a nonconstant map to the real line, so if it were connected, its image would be an interval with cardinality c. Thus connected Urysohn spaces have cardinality $\geq c$. Urysohn [127] showed that connected regular spaces must be uncountable (a separation of a countable regular space can be constructed by induction), and also that there exist countable connected Hausdorff spaces.

Now the absence of any nonconstant continuous real-valued functions is a very strong form of connectedness which cannot occur in Urysohn spaces; we will call such spaces **strongly connected.** Hewitt's example shows that regular spaces may be strongly connected. Figure 31 indicates the relations between these concepts, with the designation of certain significant counterexamples. For simplicity here, we assume the continuum hypothesis.

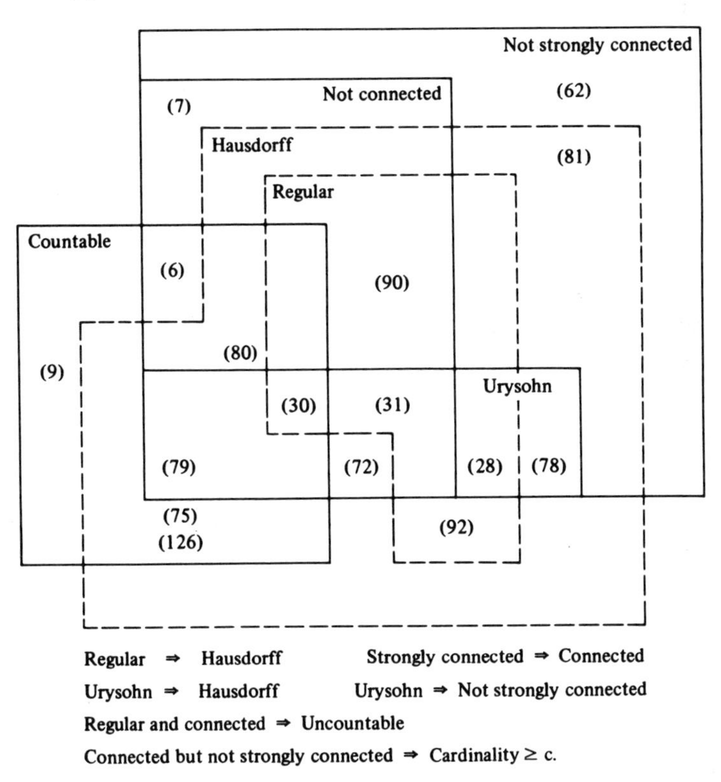

Figure 31.

Van Douwen [130] has given an analogous construction of a regular space on which every continuous real valued function is constant. His example makes particularly clear use of twins (pairs of points which continuous functions cannot separate) and of the condensation process whereby a space is constructed from infinitely many spaces containing twins in such a way as to permit only constant functions.

93. This is an adaptation of an example given by Thomas [123]; by translating his example into planks and corkscrews, we hope to make clear the similarity between Thomas' example and Tychonoff's.

95–97. Each of these examples is from Alexandroff and Urysohn [5], pp. 6, 13.

98. This example is discussed by Appert in [10]; it was introduced by Urysohn [127].

99–100. Both of these spaces are adaptations of examples developed by Ramanathan in [97] and [98].

101. From Alexandroff and Urysohn [5], p. 15.

103. The proof in 103.3 that X_λ is separable whenever $\lambda \leq 2^{\aleph_0}$ is a special application of a proof in Dugundji [35], p. 175 that separability is preserved under products of cardinality $\leq 2^{\aleph_0}$. The proof in 103.6 that X_λ is not normal is adapted from Stone [115].

107. The treatment of Helly space is motivated by a problem in Kelly [64], p. 164.

109. Early basic results concerning box topologies are discussed in Knight [67]. M. E. Rudin [109] has shown that the continuum hypothesis implies that R^ω is normal since it is a box product of countably many locally compact metrizable spaces. On the other hand Van Douwen [131] has shown that the box product topology need not, in general, be normal by the following example. Let $X_i = \{0\} \cup \left\{\frac{1}{n} \,\middle|\, n \in Z^+\right\}$ and let Y be the irrationals. Then in the box topology $Y \times \prod_{i=1}^{\infty} X_i$ is not normal. In fact, Van Douwen proves that a box product of metrizable spaces cannot be completely normal if infinitely many factors

are non-discrete. Further details and a summary of recent literature may be found in [107].

110. Most of the material about βX in its ultrafilter guise is adapted from Gillman and Jerison [43], Chapters 6 and 9. The original ideas are primarily due to Stone [116]. The method in Čech [28] is similar to the view of βX as a subspace of a product space. The characterizing and mapping properties of βX described in 110.3 express the fact that β is a functor from the category of completely regular spaces to itself whose range is compact Hausdorff spaces, and that h is a natural transformation from the identity to β.

111. This discussion in 111.4 is from Gillman and Jerison [43]. In particular the claim that B_s satisfies the finite intersection property is proved there in detail. In recent years the Stone-Čech compactification has continued to be a useful source of examples. In [135] Warren shows that for the integers N, βN-N-$\{p\}$ is not normal for any $p \in \beta N$-N. Further uses of βN are mentioned below in Note 130.

112. This example is due to J. Novak [90].

113. Notice that in this topology the closure of a basis set is the same as its closure in the Stone-Čech compactification. As with the various dense extension topologies this renders the space not T_3. This is a frequent occurrence that in an expanded topology the closures of open sets remain unchanged.

119. The special infinite broom of 119.4 is adapted from Hocking and Young [52], p. 113 where a similar space is used to illustrate the concept of **connected im kleinen.**

124. The construction in this example is due to Hocking and Young [52], p. 110 who modified an idea of Bernstein [17]. Two aspects of this example are worth further comment. In the construction we assume that every nondegenerate closed connected subset of the plane has the cardinality c of the continuum. To see this we merely consider the image of such sets under nonconstant continuous maps to the real line. Such images must be connected, so must be nondegenerate intervals which have cardinality c. Thus the original connected closed subset has cardinality c.

Secondly, in 124.3, we use the obvious but subtle fact that every set which separates the plane must contain a nondegenerate closed connected set. This follows directly from the theorem (see, for instance, Newman [88], p. 124) that every component of the complement of a connected open subset of the plane has a connected boundary. The complement of the union of two disjoint open subsets of the plane will always contain some such boundary.

125. Further details, particularly concerning 125.4, can be found in Gustin [46].

126. This example comes from Roy [102] who seems to have started an active line of investigation. In [80] Miller presents two related examples, one a Urysohn space with a dispersion point, the other a locally connected space with dispersion point. A later example of Ritter [101] has similar properties. About two years later both Vought [134] and Kannan and Rajagopalen [62] showed just how far one can go. A countable connected space cannot be regular, but both papers present examples of countable Urysohn spaces which are regular almost everywhere (regular at a dense set of points). In fact Kannan and Rajagopalen show that there are 2^c such spaces.

Baggs in [15] treats the question of maximal connected topologies for a space. He modifies Roy's basic construction to produce a connected Hausdorff space which cannot be a subspace of any maximal connected space.

128. This example is from Knaster and Kuratowski [66], p. 241, a paper which contains many similar examples.

130. The history of indecomposable continua may be traced back to Janiszewski [53] whose example is rather different from the one presented here. Our example is presented by Bing [19] though we call it a pseudo-arc which was the term used by Moise [81], for a different description of the same space. The footnotes in Moise and Miller [79] give a good survey of the basic literature dealing with the unusual properties of indecomposable continua. More recently Bellamy [16] has used the Stone-Čech compactification (Example 110) to give a new example of an indecomposable continuum: if $A = [1, \infty)$ (with the Euclidean topology), then $\beta(A) - A$ is an indecomposable continuum.

131. For the extended proof that such an inductive construction can be performed one should see Miller [79]. It appears that the indecomposable continuum K of this example may be constructed in the spirit of Bing [19] and Moise [81]. It is necessary to keep

certain links of the chains long, but thin, and to adapt the arguments of Bing [19] to use arguments about adjacent pairs of chains as used by Moise to prove such a continuum is indecomposable. Our proofs that X is biconnected and has no dispersion point are adopted from Miller.

133. This example was constructed by Tangora [121] as the solution to a Monthly problem.

136. This is adapted from Duncan [36].

141. This is one of many examples in Bing [22].

142. The space (X,σ) is also from Bing [22] who introduced it as an example of a normal space which is not collectionwise normal. Michael [77] selected the subspace Y to be metacompact. That normal metacompact spaces are countably paracompact was proved by Morita [84]. These papers discuss at length several areas beyond the scope of this book, all related to the metrization problem. An introductory survey of significant recent results in metrization theory appears above in Part III. A more advanced and more recent survey can be found in [107].

BIBLIOGRAPHY

[1] Alexandroff, P. S. General duality theorems for non-closed sets of n-dimensional space. *Mat. Sbornik* **21** (63) (1947) 161–231.

[2] Alexandroff, P. S. On the metrization of topological spaces. *Bull Acad. Polon. Sci.* **8** (1960) 135–140.

[3] Alexandroff, P. S. Some results in the theory of topological spaces, obtained within the last twenty-five years. *Russian Math. Surveys* **15** (1960) 23–83.

[4] Alexandroff, P. S. Sur les ensembles de la première classe et les espaces abstraits. *Comptes Rendus Acad. Sci. Paris* **178** (1924) 185.

[5] Alexandroff, P. S. and Urysohn, P. Mémoire sur les espaces topologiques compacts. *Verh. Konink. Acad. Wetensch. Amsterdam* **14** (1929) 1–96.

[6] Alexandroff, P. S. and Urysohn, P. Une condition nécessaire et suffisante pour qu'une classe (*L*) soit une classe (*B*). *Comptes Rendus Acad. Sci. Paris* **177** (1923) 1274–1276.

[7] Alster, K. and Przymusiński, T. Normality and Martin's axiom. *Fund Math.* **91** (1976) 123–131.

[8] Anderson, R. D. Hilbert space is homeomorphic to the countable infinite product of lines. *Bull. Amer. Math. Soc.* **72** (1966) 515–519.

[9] Anderson, R. D. and Bing, R. H. A complete elementary proof that Hilbert space is homeomorphic to the countable infinite product of lines. *Bull. Amer. Math. Soc.* **74** (1968) 771–792.

[10] Appert, Q. *Propriétés des Espaces Abstraits les Plus Généraux.* Actual. Sci. Ind. No. 146, Herman, 1934.

[11] Arens, R. Note on convergence in topology. *Math. Mag.* **23** (1950) 229–234.

[12] Arhangel'skiĭ, A. Bicompact sets and the topology of spaces. *Soviet Math. Dokl.* **4** (1963) 561–564.

[13] Arhangel'skiĭ, A. Mappings and spaces. *Russian Math. Surveys* **21** (1966) 115–162.

[14] Arhangel'skiĭ, A. On the metrization of topological spaces. *Bull. Acad. Polon. Sci.* **8** (1960) 589–595.

[15] Baggs, I. A connected Hausdorff space which is not contained in a maximal connected space. *Pacific J. Math.* **51** (1974) 11–18.

[16] Bellamy, D. P. A non-metric indecomposable continuum. *Duke J. of Math.* **38** (1971) 15–20.

[17] Bernstein, F. Zur theorie der trigonometrischen reihe. *Leipzig Bericht* **60** (1908) 329.

[18] Bing, R. H. A connected, countable Hausdorff space. *Proc. Amer. Math. Soc.* **4** (1953) 474.

[19] Bing, R. H. A homogeneous indecomposable plane continuum. *Duke Math. J.* **15** (1948) 729–742.

[20] Bing, R. H. A translation of the normal Moore space conjecture. *Proc. Amer. Math. Soc.* **16** (1965) 612–619.

[21] Bing, R. H. Challenging conjectures. *Amer. Math. Monthly* (50th Anniv. Issue) **74** (1967) 56–64.

[22] Bing, R. H. Metrization of topological spaces. *Canad. J. Math.* **3** (1951) 175–186.

[23] Borges, C. R. On the metrizability of topological spaces. *Canad. J. Math.* **20** (1968) 795–804.

[24] Borges, C. R. Stratifiable spaces. *Pacific J. Math.* **17** (1966) 1–16.

[25] Bourbaki, N. *Elements of Mathematics: General Topology.* Addison-Wesley, Reading, Mass., 1966.

[26] Brown, M. Semi-metric spaces. *Summer Institute on Set Theoretic Topology (Madison).* Amer. Math. Soc., Providence, 1955, pp. 64–66.

[27] Bukovsky, L. Borel subsets of metric separable spaces. *General Topology and its Relations to Modern Analysis and Algebra.* Academic Press, New York, 1967, pp. 83–86.

[28] Čech, E. On bicompact space. *Ann. of Math.* **38** (1937) 823–844.

[29] Cech, E. Sur la dimension des espaces parfaitement normaux. *Bull. Intern. de l'Acad. de Bohême* (Prague) **33** (1932) 38–55.

[30] Ceder, J. G. Some generalizations of metric spaces. *Pacific J. Math.* **11** (1961) 105–125.

[31] Chittenden, E. W. On the metrization problem and related prob-

lems in the theory of abstract sets. *Bull. Amer. Math. Soc.* **33** (1927) 13–34.

[32] Cohen, P. J. The independence of the continuum hypothesis, I, II. *Proc. Nat. Acad. Sci.* **50** (1963) 1143–1148; **51** (1964) 105–110.

[33] Dieudonné, J. Une généralisation des espaces compacts. *J. Math. Pure Appl.* **23** (1944) 65–76.

[34] Dowker, C. H. On countably paracompact spaces. *Canad. J. Math.* **3** (1951) 219–224.

[35] Dugundji, J. *Topology.* Allyn and Bacon, Boston, Mass., 1966.

[36] Duncan, R. L. A topology for the sequence of integers. *Amer. Math. Monthly* **66** (1959) 34–39.

[37] Engelking, R. *Outline of General Topology.* North-Holland, Amsterdam, 1968.

[38] Fleissner, W. G. A normal collectionwise Hausdorff, not collectionwise normal space. *General Topology and Appl.* **6** (1976) 57–64.

[39] Fleissner, W. G. Normal Moore spaces in the constructible universe. *Proc. Amer. Math. Soc.* **46** (1974) 294–298.

[40] Fleissner, W. G. When is Jones' space normal? *Proc. Amer. Math. Soc.* **50** (1975) 375–378.

[41] Fort, M. K., Jr. Nested neighborhoods in Hausdorff spaces. *Amer. Math. Monthly* **62** (1955) 372.

[42] Gaal, S. A. *Point Set Topology.* Academic Press, New York, 1964.

[43] Gillman, L. and Jerison, M. *Rings of Continuous Functions.* Van Nostrand Reinhold Press, Princeton, New Jersey, 1960; Springer-Verlag, New York, 1976.

[44] Gödel, K. *The Consistency of the Continuum Hypothesis.* Princeton Univ. Press, Princeton, New Jersey, 1940.

[45] Greever, J. *Theory and Examples of Point Set Topology.* Brooks-Cole, Belmont, Calif., 1967.

[46] Gustin, W. Countable connected spaces. *Bull. Amer. Math. Soc.* **52** (1946) 101–106.

[47] Heath, R. W. On certain first countable spaces. *Topology Seminar.* Wisconsin Univ. Press, 1965; Princeton Univ. Press, 1966, pp. 103–113.

[48] Heath, R. W. On spaces with point-countable bases. *Bull. Acad. Polon. Sci.* **13** (1965) 393–395.

[49] Heath, R. W. Screenability, pointwise paracompactness, and metrization of Moore spaces. *Canad. J. Math.* **16** (1964) 763–770.

[50] Heath, R. W. Separability and $\aleph_1$ compactness. *Colloq. Math.* **12** (1964) 11–14.

[51] Hewitt, E. On two problems of Urysohn. *Annals of Math.* **47** (1946) 503–509.

[52] Hocking, J. G. and Young, G. S. *Topology*. Addison-Wesley, Reading, Mass., 1961.

[53] Janiszewski, Z. Zur les continus irréductible entre deux points. *J. de l'Ecole Polytechnique* **16** (1912) 114.

[54] Jech, T. Non-provability of Souslin's hypothesis. *Comm. Math. Univer. Carolinae* **8** (1967) 291–305.

[55] Jensen, R. B. The fine structure of the constructible universe. *Ann. Math. Logic* **4** (1972) 229–308.

[56] Jones, F. B. Concerning normal and completely normal spaces. *Bull. Amer. Math. Soc.* **43** (1937) 671–677.

[57] Jones, F. B. Metrization. *Amer. Math. Monthly* **73** (1966) 571–576.

[58] Jones, F. B. Moore spaces and uniform spaces. *Proc. Amer. Math. Soc.* **9** (1958) 483–486.

[59] Jones, F. B. Remarks on the normal Moore space metrization problem. *Topology Seminar.* Wisconsin Univ. Press, 1965; Princeton Univ. Press, 1966, pp. 115–119.

[60] Juhász, I. *Cardinal Functions in Topology.* Math. Centre Tracts 34, Mathematical Centre, Amsterdam, 1971.

[61] Juhász, I. Martin's axiom solves Ponomarev's problem. *Bull. Acad. Polon. Sci.* **18** (1970) 71–74.

[62] Kannan, V. and Rajagopalan, M. Regularity and dispersion in countable spaces. *Duke J. Math.* **39** (1972) 729–734.

[63] Kaplan, S. Homology properties of arbitrary subsets of Euclidean spaces. *Trans. Amer. Math. Soc.* **62** (1947) 248–271.

[64] Kelley, J. L. *General Topology*. Van Nostrand Reinhold, Princeton, New Jersey, 1955; Springer-Verlag, New York, 1975.

[65] Kirch, A. M. A countable connected locally connected Hausdorff space. *Amer. Math. Monthly* **76** (1969) 169.

[66] Knaster, B. and Kuratowski, C. Sur les ensembles connexes. *Fund. Math.* **2** (1921) 206–255.

[67] Knight, C. J. Box topologies. *Quart. J. Math. Oxford Ser.* (2) **15** (1964) 41–54.

[68] Kuratowski, C. *Topologie I.* Monografie Matematyczne Vol. 20, Warsaw, 1958.

[69] Mansfield, M. J. On countably paracompact normal spaces. *Canad. J. Math.* **9** (1957) 443–449.

[70] Martin, D. A. and Solovay, R. M. Internal Cohen extensions. *Ann. Math. Logic* **2** (1970) 143–178.

[71] Mazurkiewicz, S. Sur les continus indécomposables. *Fund. Math.* **10** (1927) 305–310.

[72] McAuley, L. F. A note on complete collectionwise normality and paracompactness. *Proc. Amer. Math. Soc.* **9** (1958) 796–799.

[73] McAuley, L. F. A relation between perfect separability, completeness, and normality in semimetric spaces. *Pacific J. Math.* **6** (1956) 315–326.

[74] McAuley, L. F. Paracompactness and an example due to F. B. Jones. *Proc. Amer. Math. Soc.* **7** (1956) 1155–1156.

[75] Michael, E. A note on paracompact spaces. *Proc. Amer. Math. Soc.* **4** (1953) 831–838.

[76] Michael, E. $\aleph_0$ spaces. *J. Math. Mech.* **15** (1966) 983–1002.

[77] Michael, E. Point finite and locally finite coverings. *Canad. J. Math.* **7** (1955) 275–279.

[78] Michael, E. The product of a normal space and a metric space need not be normal. *Bull. Amer. Math. Soc.* **69** (1963) 375–376.

[79] Miller, E. W. Concerning biconnected sets. *Fund. Math.* **29** (1937) 123–133.

[80] Miller, G. G. Countable connected spaces. *Proc. Amer. Math. Soc.* **26** (1970) 355–360.

[81] Moise, E. E. An indecomposable plane continuum which is homeomorphic to each of its non-degenerate subcontinua. *Trans. Amer. Math. Soc.* **63** (1948) 581–594.

[82] Moore, R. L. *Foundations of Point Set Theory*. Colloq. Publ. No. 13, Amer. Math. Soc., New York, 1932.

[83] Morita, K. Products of normal spaces with metric spaces. *Math. Ann.* **154** (1964) 365–382.

[84] Morita, K. Star-finite coverings and the star-finite property. *Math. Japon.* **1** (1948) 60–68.

[85] Murdeshwar, M. G. and Naimpally, S. A. *Quasi-Uniform Topological Spaces*. Noordhoff, Groninger, Netherlands, 1966.

[86] Nagami, K. Paracompactness and strong screenability. *Nagoya Math. J.* **8** (1955) 83–88.

[87] Nagata, J. On a necessary and sufficient condition of metrizability. *J. Inst. Polytech. (Osaka)* **1** (1950) 93–100.

[88] Newman, M. H. A. *Elements of the Topology of Plane Sets of Points*. Cambridge Univ. Press, New York, 1939.

[89] Niemytzki, V. and Tychonoff, A. Beweis des Satzes, dass ein metrisierbarer Raum dann and nur dann kompakt ist, wenn er in jeder Metrik vollständig ist. *Fund. Math.* **12** (1928) 118–120.

[90] Novak, J. On the Cartesian product of two compact spaces. *Fund. Math.* **40** (1953) 106–112.

[91] Ostaszewski, A. J. A countably compact, first-countable, hereditarily separable regular space which is not completely regular. *Bull. Acad. Polon. Sci.* **23** (1975) 431–435.

[92] Pervin, W. J. *Foundations of General Topology*. Academic Press, New York, 1964.

[93] Pixley, C. and Roy, P. Uncompletable Moore spaces. *Topology Conference*. Auburn University, 1969, pp. 75–85.

[94] Przymusiński, T. A Lindelöf space X such that X^2 is normal but not paracompact. *Fund. Math.* **78** (1973) 291–296.

[95] Przymusiński, T. A note on collectionwise normality and product spaces. *Colloq. Math.* **23** (1975) 65–70.

[96] Przymusiński, T. and Tall, F. The undecidability of the existence of a nonseparable normal Moore space satisfying the countable chain condition. *Fund. Math.* **85** (1974) 291–297.

[97] Ramanathan, A. Maximal-Hausdorff space. *Proc. Ind. Acad. Sci.* **26** (1947) 31–42.

[98] Ramanathan, A. Minimal bicompact spaces. *J. Indian Math. Soc.* **12** (1948) 40–46.

[99] Reed, G. M. (Ed). *Set-Theoretic Topology*. Academic Press, New York, 1977.

[100] Reed, G. M. and Zenor, P. Metrization of Moore spaces and generalized manifolds. *Fund. Math.* **91** (1976) 203–210.

[101] Ritter, G. X. A connected, locally connected, countable Hausdorff space. *Amer. Math. Monthly* **83** (1976) 185–186.

[102] Roy, P. A countable connected Urysohn space with a dispersion point. *Duke Math. J.* **33** (1966) 331–333.

[103] Rudin, M. E. A normal hereditarily separable non-Lindelöf space. *Ill. J. Math.* **16** (1972) 621–626.

[104] Rudin, M. E. A normal space X for which $X \times I$ is not normal. *Fund. Math.* **73** (1971) 179–186.

[105] Rudin, M. E. A separable normal nonparacompact space. *Proc. Amer. Math. Soc.* **7** (1956) 940–941.

[106] Rudin, M. E. Countable paracompactness and Souslin's problem. *Canad. J. Math.* **7** (1955) 543–547.

[107] Rudin, M. E. *Lectures on Set Theoretic Topology*. CBMS Regional Conference Series No. 23, Amer. Math. Soc., Providence, 1975.

[108] Rudin, M. E. Souslin's conjecture. *Amer. Math. Monthly* **76** (1969) 1113–1119.

[109] Rudin, M. E. The box product of countably many compact metric spaces. *Gen. Top.* **2** (1972) 293–298.

[110] Smirnov, Yu. M. On metrization of topological spaces. *Uspekhi Matem. Nauk.* **6** (1951) 100–111; *Amer. Math. Soc. Transl.* No. **91.**

[111] Solovay, R. and Tennenbaum, S. Iterated Cohen extensions and Souslin's problem. *Ann. of Math.* **94** (1971) 201–245.

[112] Sorgenfrey, R. H. On the topological product of paracompact spaces. *Bull. Amer. Math. Soc.* **53** (1947) 631–632.

[113] Souslin, M. Problème 3. *Fund. Math.* **1** (1920) 223.

[114] Steen, L. A direct proof that a linearly ordered space is hereditarily collectionwise normal. *Proc. Amer. Math. Soc.* **24** (1970) 727–728.

[115] Stone, A. H. Paracompactness and product spaces. *Bull. Amer. Math. Soc.* **54** (1948) 977–982.

[116] Stone, M. H. Applications of the theory of Boolean rings to general topology. *Trans. Amer. Math. Soc.* **41** (1937) 375–481.

[117] Tall, Franklin D. A counterexample in the theories of compactness and of metrization. *Indag. Math.* **35** (1973) 471–474.

[118] Tall, Franklin D. New results on the normal Moore space problem. *Proc. of Washington State Univ. Conf. on General Topology.* Washington State Univ., 1970, pp. 120–126.

[119] Tall, Franklin D. On the existence of normal metacompact Moore spaces which are not metrizable. *Canad. J. Math.* **26** (1974) 1–6.

[120] Tamano, H. Normality and product spaces. *General Topology and its Relations to Modern Analysis and Algebra.* Academic Press, 1967, pp. 349–352.

[121] Tangora, M. Connected unions of disconnected sets. *Amer. Math. Monthly* **72** (1965) 1038.

[122] Tennenbaum, S. Souslin's problem. *Proc. Nat. Acad. Sci.* **59** (1968) 60–63.

[123] Thomas, J. A regular space, not completely regular. *Amer. Math. Monthly* **76** (1969) 181.

[124] Traylor, D. R. On normality, pointwise paracompactness and the metrization question. *Topology Conference.* Arizona State Univ. 1967, pp. 286–292.

[125] Tukey, J. W. *Convergence and Uniformity in Topology.* Princeton Univ. Press, 1940.

[126] Tychonoff, A. Über einen Metrisationssatz von P. Urysohn. *Math. Ann.* **95** (1926) 139–142.

[127] Urysohn, P. Uber die Machtigkeit der zusammenhängenden Mengen. *Math. Ann.* **94** (1925) 262–295.

[128] Urysohn, P. Uber Metrization des kompakten topologischen Raumes. *Math. Ann.* **92** (1924) 275–293.

[129] Urysohn, P. Zum Metrizations problem. *Math. Ann.* **94** (1925) 309–315.

[130] Van Douwen, E. K. A regular space on which every continuous real-valued function is constant. *Nieuw Archief voor Wiskunde* **20** (1972) 143–145.

[131] Van Douwen, E. K. The box product of countably many metrizable spaces need not be normal. *Fund. Math.* **88** (1975) 127–132.

[132] Van Douwen, E. K. The Pixley-Roy topology on spaces of subsets.

Set-Theoretic Topology. Academic Press, New York, 1977, pp. 111–134.

[133] Vickery, C. W. Axioms for Moore spaces and metric spaces. *Bull. Amer. Math. Soc.* **46** (1940) 560–564.

[134] Vought, E. J. A countable connected Urysohn space with a dispersion point that is regular almost everywhere. *Colloq. Math.* **28** (1973) 205–209.

[135] Warren, N. M. Properties of Stone-Čech compactifications of discrete spaces. *Proc. Amer. Math. Soc.* **33** (1972) 599–606.

[136] Weil, A. *Sur les Espaces à Structure Uniforme*. Actual. Sci. Ind. No. 551, Hermann, 1937.

[137] Whyburn, G. T. *Analytical Topology*. Colloq. Publ. No. 28, Amer. Math. Soc., Providence, 1942.

[138] Worrell, J. M., Jr. and Wicke, H. H. Characterizations of developable topological spaces. *Canad. J. Math.* **17** (1965) 820–830.

[139] Younglove, J. N. Two conjectures in point set theory. *Topology Seminar*. Wisconsin Univ. Press, 1965; Princeton Univ. Press, 1966, pp. 121–123.

Index